T0230239

Lecture Notes in Artificial Intelligence 1177

Subseries of Lecture Notes in Computer Science
Edited by J. G. Carbonell and J. Siekmann

Lecture Notes in Computer Science

Edited by G. Goos, J. Hartmanis and J. van Leeuwen

Springer

Berlin
Heidelberg
New York
Barcelona
Budapest
Hong Kong
London
Milan
Paris
Santa Clara
Singapore
Tokyo

Jörg P. Müller

The Design of Intelligent Agents

A Layered Approach

 Springer

Series Editors
Jaime G. Carbonell, Carnegie Mellon University, Pittsburgh, PA, USA
Jörg Siekmann, Universität des Saarlandes, Saarbrücken, Germany

Author

Jörg P. Müller
Mitsubishi Electric Digital Library Group
103 New Oxford Street, London WC1A 1EB, U.K.
E-mail: jpm@dlib.com

Cataloging-in-Publication Data applied for

Die Deutsche Bibliothek - CIP-Einheitsaufnahme

Müller, Jörg P.:
The design of intelligent agents : a layered approach / Jörg P.
Müller. - Berlin ; Heidelberg ; New York ; Barcelona ;
Budapest ; Hong Kong ; London ; Milan ; Paris ; Santa Clara ;
Singapore ; Tokyo : Springer, 1996
  (Lecture notes in computer science ; Vol. 1177 : Lecture notes in
  artificial intelligence)
  ISBN 3-540-62003-6
NE: GT

CR Subject Classification (1991): I.2.11, I.2.9, I.2, D.2

ISBN 3-540-62003-6 Springer-Verlag Berlin Heidelberg New York

© Springer-Verlag Berlin Heidelberg 1996
Printed in Germany

Typesetting: Camera ready by author
SPIN 10550188   06/3142 – 5 4 3 2 1 0   Printed on acid-free paper

Foreword

Open any introductory textbook on programming, and you will be presented with the following axiom: computers do exactly what they are told, and nothing else. This is such a fundamental aspect of our relationship with computers that we rarely stop to think about it. And, for the most part, it doesn't bother us — we are happy to have such obedient servants, even if they do lack imagination. However, for many important applications, obedience is not enough: we need systems that can decide *for themselves* what they need to do in any given situation. Consider a space probe on a mission to the outer planets. Such a probe *cannot* be controlled in real time — the laws of physics forbid it. Nor can the probe's mission be entirely pre-planned: in all but the shortest of space missions, unforeseen events are inevitable. The probe must therefore be capable of *autonomous action* in order to achieve its design objectives. In short, it must decide for itself what to do in the event of unforeseen circumstances. Moreover, it cannot spend forever deciding what to do: decisions must be made in time for them to be useful.

While this example is perhaps somewhat extreme, the requirement for autonomy is not. It is now recognised that autonomous components are the most appropriate solution for many important software problems. To pick a topical example, programs traversing the INTERNET in search of some information are subject to the same basic autonomy requirement as our space probe. It is neither desirable nor feasible to directly control such programs, and the dynamic nature of the INTERNET makes it impossible to pre-plan a search with any confidence.

Computer systems capable of flexible autonomous action are known as *agents*. Like so many other problems in AI, developing agents is a much harder problem than it might at first appear. In particular, integrating *goal-driven* and *event-driven* behaviours in a convincing, clean *agent architecture* is a rich, challenging area of ongoing research. Designing agents that can efficiently *cooperate* with other agents in order to achieve their goals is a similarly important area, still somewhat in its infancy. This book represents a significant advance in the design of cooperative autonomous agents. It describes an important new agent architecture called INTERRAP. The concepts underpinning INTERRAP, the data structures it contains, the operations that manipulate these structures, and the application of the architecture are all described in careful, unambiguous detail, making the transfer of techniques from INTERRAP to other architectures a realistic possibility.

Autonomous agents have the potential to play an important role in complex distributed systems of the twenty-first century. A scientific, engineering approach to agent design is essential if this potential is to be realised; this book represents just such an approach. For this reason, *The Design of Intelligent Agents* represents a real advance in application-oriented research on intelligent agents. It is essential reading for anyone wanting to understand and apply this exciting new software technology.

September 1996   Michael Wooldridge

Preface

I am indebted to Jörg Siekmann who was the supervisor of my dissertation on which this book is based. Through his way of looking at the world and through his never-ending enthusiasm, he raised my passion for Artificial Intelligence when I was an undergraduate on his introductory AI courses, and he has kept this fire burning ever since.

Alfred Hofmann encouraged me to publish my PhD thesis as a book and gave me cooperative and reliable guidance and support through the whole process of publication.

I would like to thank all of my colleagues at the DFKI, in particular those in the projects AKA-MOD, CoMMA-MAPS, KIK-TEAMWARE, and TACOS, who accompanied me through this work over the past four years. A considerable part of the research described in this book was done while I was a member of the AKA-MOD research project. I thank Klaus Fischer, Norbert Kuhn, Jürgen Müller, Andreas Schroth, Achim Schupeta, and Thomas Weiser for their support and for many—sometimes controversial—discussions without which this work would not have reached its current state. Alastair Burt, Michael Kolb, Andreas Lux, and Donald Steiner were always cooperative colleagues. Special thanks also to Manfred Meyer who was the supervisor of my master's thesis and who introduced me to the art of writing papers that are (at least sometimes) accepted.

I am greatly indebted to Markus Pischel for having been a constant source of inspiration and for so often having great ideas of what to do next. The main concepts of INTERRAP as well as the design and implementation of the FORKS simulation system were elaborated in close collaboration with him. Over and above, he has not only been a great colleague, but also an honest and reliable friend.

Michael Rosinus, Erwin Margewitsch, Wolfgang Morell, Michael Niehren, and Jürgen Scherer contributed to the implementation of the FORKS system.

Michael Beetz, Hans-Dieter Burkhard, Klaus Fischer, Markus Pischel, Jörg Siekmann, and Mike Wooldridge read draft versions of this dissertation or parts of it. Their suggestions and constructive criticism have contributed to its present form. No need to say that I am fully responsible for any remaining flaws.

During my time at the DFKI, I have had the great opportunity to meet many people doing research in DAI. I would like to thank them for being such an extraordinary research community and for being open to my ideas.

I thank Robert Kowalski, who was the scientific mentor of the research projects AKA-MOD and CoMMA-MAPS at the DFKI, for his interest in my work and for fruitful discussions about reactivity and deliberation; Mike Wooldridge for his encouragement; Mike Georgeff and Anand Rao for clarifying comments on BDI architectures; Martin Buchheit for his constant supply of coffee milk; Gil Tidhar for discussions about multiagent planning; Gilad Zlotkin for great conversations about game theory, negotiation, and their relationship to world politics; Innes Ferguson for chats about agent architectures and for a great weekend in Ottawa; Bernhard Hollunder for thrilling squash games; Nick Jennings for asking me each time we met—not without a hint of doubt—when I would finally finish my PhD; Afsaneh Haddadi for her moral support at an intermediate stage of this dissertation, and for making the point that I did not have to put into it *everything* I had done so far.

My family has always been a place for me to recover and recharge my batteries. My mother, and my aunt Anneliese and uncle Kurt have contributed so much to what I am today and have given me the strength to bring this work to a good end.

Eva, thanks for your love and great support—this work is dedicated to you!

September 1996   Jörg Müller

Table of Contents

List of Figures

*So wohl es getan ist, seinen Plan im ganzen gehörig zu überlegen, so hat doch
die Ausführung, wenn sie mit der Erfindung gleichzeitig ist, so große Vorteile,
die nicht zu versäumen sind.*

JOHANN WOLFGANG VON GOETHE, AN SCHILLER, 5.6.1799.

1. Motivation

The increasing complexity of organizations and computer-controlled techni-cal processes and systems makes it impossible to design them as monolithic entities and to maintain and monitor them by centralistic control systems. Examples are the optimization of the flow of work and information through cooperating companies (*virtual enterprises*), international air traffic control, the coordination of logistic processes in shipping companies that schedule hundreds of transportation vehicles and that participate in multimodal trans-port, the use of Flexible Transport Systems (FTS) in industrial manufactur-ing and assembly, and the design and the operation of traffic guidance and control systems. The inherent distribution of competence, control, and infor-mation, as well as the complexity of the theoretical problems, e.g., scheduling problems, underlying most of these applications call for new ways and mech-anisms of domain modeling and problem-solving. Robustness has replaced optimality as the main criterion for measuring the quality of these systems. An additional requirement is *interoperability*, i.e., the ability of different het-erogeneous systems to exchange information and to work together in rapidly changing environments (e.g., in the Internet). Interoperability has to be based on open, flexible architectures.

Agent-based technologies

The rapid development of agent-based technologies since the beginning of the 1990s[1] has to be viewed in the light of these requirements. This new discipline has emerged from a multitude of parental research areas, the most important of which are symbolic Artificial Intelligence (AI), control theory, and Distributed Artificial Intelligence (DAI) (see also Chapter 2 for a detailed discussion of the background). Agents are autonomous or semi-autonomous hardware or software systems that perform tasks in complex, dynamically changing environments. Autonomy means the ability to make decisions based on an internal representation of the world, without being controlled by a cen-tral instance. Agents communicate with their environment and effect changes in their environment by executing actions. A multiagent system (MAS) con-sists of a group of agents that can take specific roles within an organiza-

---

[1] See e.g., [Mae90a], [WJ95a], [WMT96] for a survey.

tional structure. The step from isolated single-agent scenarios—that have been investigated by AI research for more than three decades—to open multiagent systems offers the new quality of *emergent behavior*: the group of agents is more than the sum of the capabilities of its members. Researchers have made use of this property to build systems for complex applications like airport management [RG95], transport logistics [FMP96], traffic guidance systems [LS95], advanced robotics systems [BKMS96], and distributed electricity management [Jen94].

Agent capabilities

In order to cope with these difficult tasks, agents need basic capabilities: firstly, they should be *reactive*, i.e., react timely and appropriately to unforeseen events and to changes in the environment. Secondly, they should be capable of *deliberation* to perform their tasks in a goal-directed manner. Thirdly, they should solve their tasks *efficiently* by making use of hard-wired procedures in routine situations. Fourthly, they have to deal with positive and negative *interactions* with other agents. In open, multiagent environments, this ability of coordination is of particular importance. Fifthly, agents need to be *adaptable* to changing environmental conditions.

The extent to which agents can meet these requirements is restricted by the limitation of their computational resources. For instance, time spent for reaction will be lacking for goal-directed reasoning and vice versa. Thus, it is a central task to define a *control architecture* for resource-bounded agents, which allows the designer of an agent-based system to integrate the requirements mentioned above, and to define the trade-offs between them in a way that is adequate for the application domain under consideration.

The thesis of this book

In this book, we describe INTERRAP[2], a pragmatic control architecture for dynamically interacting agents. Starting from the state-of-the-art approach of using a layered architecture for integrating reactive and deliberative capabilities (see Chapter 2 for examples), the message of this book is as follows:

- It is both feasible and useful to extend an architecture consisting of layers for reacting and local planning by (i) the notion of procedures that allow the agent to perform routine tasks efficiently, and by (ii) a cooperative planning layer that is located on top of the local planning layer and that enables an agent to reason about other agents and to coordinate its activities with other agents by explicitly communicating information, goals, and plans.
- Using such an architecture allows the designer of a multiagent system to deal with a broad spectrum of interactions such as conflicts among agents

---

[2] INTEGRATION OF REACTIVE BEHAVIOR AND RATIONAL PLANNING.

and collaboration to achieve a common goal, by using both local behavior-based mechanisms and mechanisms based on explicit coordination by communication.

Agent-centered vs. global perspective

This book tackles the problem of modeling agents in a multiagent system from the perspective of the designer of an individual agent. In this *agent-centered* view, an agent system is looked upon as consisting of an *agent* and its *environment*. The environment is described in the internal representation accessible to the agent in a state-based manner; the agent updates this representation by perceiving changes in the environment. Actions performed by the agent cause potentially nondeterministic transitions between states.

The agent-centered view is shared by most of the approaches to agent design discussed in Chapter 2. We believe that it is more useful for actually designing a system than the approach that takes a *global perspective* and that defines a multiagent system by a set of agents and the interactions (e.g., in terms of messages sent and received) among them. This global view, which Haddadi called *external perspective* [Had96], was adopted by Halpern and Moses [HM90] in the formal description of distributed systems. In [Bur93], Burkhard uses a similar model to prove properties of a multiagent system like liveness and fairness. The global perspective is useful for evaluating or verifying the behavior of a multiagent system. Therefore, in Chapter 5, we shall leave the "micro level" [WJ95b] of the individual agent and take the external perspective in order to evaluate the performance of INTERRAP agents in the loading dock domain.

An application example

The first application domain of the INTERRAP architecture was the modeling of an automated loading dock, where multiple autonomous forklift robots carry out transportation tasks by either loading goods onto a loading ramp, or by unloading the ramp and storing the goods on shelves. From the point of view of agent design, this application is appealing because on the one hand, it is easy to understand, while on the other hand, it offers a variety of interesting problems. The local planning of tasks and the necessity to deal with changes caused by the actions of other agents allow us to study the relationship between reactivity and deliberation. The agents in the loading dock have to react in real time, they are resource-bounded, and have incomplete knowledge about the world.

However, the main focus of this book is not on the reconciliation of reactivity and deliberation, but on the integration of agent interaction in a planner–reactor architecture. The loading dock domain allows us to investigate of various forms of agent interaction using a variety of interaction

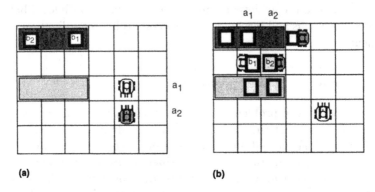

Fig. 1.1. Multiagent conflicts in the loading dock

mechanisms. For example, Figure 1.1 shows two typical situations: in Figure (1.1.a), two agents block each other in a wide hallway. Figure (1.1.b) presents a similar situation; this time, however, both agents are in a narrow corridor between two shelves, searching for free space to store boxes. Both situations are examples of *multiagent conflicts*. Obviously, there are different alternative ways of resolving them. For example, the agents could—each by itself—locally decide to dodge, e.g., by some random movement; alternatively, the agents could communicate their goals, and agree on a *joint plan* to resolve the conflict. Intuitively, the former option seems to be more adequate to deal with the situation in Figure (1.1.a), whereas the latter alternative seems to be more reasonable in Figure (1.1.b). But is this intuition correct? How will a system consisting of a number of forklifts behave if the individual agents use these or similar strategies? And, assuming there is an answer to which interaction strategy to prefer in a certain situation, how is the recognition and the handling of conflicts actually *implemented* within a layered agent architecture? These are central questions that we shall investigate in this book.

Overview

This book is structured as follows:

Chapter 2 puts the work presented throughout this book into a broader perspective. We present a brief history of the field and a summary of some important parental disciplines, before we survey the state of the art in agent design. Related work is discussed and evaluated with respect to its scope and limitations.

Chapter 3 describes the INTERRAP agent architecture. The underlying conceptual model is outlined. The structure of the architecture is depicted, its main modules, i.e., the layered control unit, the knowledge base, and the world interface, are described, and an operational model for the individual control layers and their interplay is given.

Chapter 4 illustrates the general agent model by presenting a specific application: the simulation of an automated loading dock with autonomous forklift robots. It is shown how these agents are modeled using INTER-RAP. Alternatives in the design of agent interaction are pointed out by the example of handling multiagent conflicts in a decentralized manner. Examples are given both for local, behavior-based and for deliberative, communication-based interaction mechanisms.

Chapter 5 evaluates the work presented in this book. It focuses on the effects of using different methods for the local design of agents and their interactions in the loading dock domain on the behavior and the efficiency of the multiagent system as a whole. A probabilistic model is presented that allows the agent designer to predict the global effects of a specific class of behavior-based decision algorithms. It is complemented by results of a series of experiments that were carried out using the FORKS simulation system, and that compare the behavior of agents using behavior-based algorithms for conflict resolution with those relying on coordination through communication.

Chapter 6 summarizes the results of this book. Problems that go beyond the scope of this work as well as areas and topics for future research are identified.

2. Background

2.1 Introduction

Control architectures for embedded systems are by no means an invention of computer scientists or AI researchers. Their philosophical roots date back to Aristotle, and their systematic investigation started in the 18th century when James Watt laid the foundation of modern control theory by proposing the use of mechanical feedback to control steam engines. In the nineteen forties and fifties, promoted by the availability of analog computers, the emerging field of cybernetics tried to unify the phenomena of control and communication observed in animals and machines into a common mathematical model, which soon outgrew its analogue roots. The availability of digital computers and a common agreement on Herbert Simon's *physical symbol system hypothesis* (i.e., the assumption that the ability to symbolically represent aspects of the world is a prerequisite for all intelligent behavior, [NS76]) then gave rise to the research area of AI with the Dartmouth Conference in 1956. The notion of an agent has received a central role in the research in Distributed Artificial Intelligence (DAI) [BG88], where the behavior of groups of intelligent systems is studied from different angles. While the somewhat fuzzy notion of DAI encompasses many research areas not all of which are the central domain of AI, a central area are multiagent systems (MAS) [DR94][1]. Moreover, the past five years have witnessed an explosion of interest in research on the design and the implementation of *intelligent agents* (see e.g., [WJ95a] [WMT96] for comprehensive collections of papers). This broadened the rather narrow cooperation-oriented view of MAS to more general issues of intelligent resource-bounded agents that make decisions, interact with other agents, and act autonomously and rationally in time-constrained, open, multiagent environments.

Before providing some background of the research on intelligent agent design, we shall briefly comment on the term *agent*. The term is used to characterize very different kinds of systems, starting from primitive biological systems (cells, bees, ants) over complex technological artifacts such as mobile

---

[1] In fact, it seems that DAI is by now almost identified with MAS. E.g., the ICMAS conference that has unified world-wide research on DAI was first held in 1995 and was named *International Conference on Multiagent Systems*.

robots or air planes, to systems that simulate or describe whole human societies or organizations such as shipping companies or industrial enterprises.
A definition which is generally agreed upon is not yet in sight. At the 1994
DAI Workshop, Carl Hewitt commented ironically that "asking the question
of what an agent is to a DAI researcher is as embarrassing as the question of
what intelligence means is for an AI researcher".

At this stage, we give a very reductionist delineation of what an agent is *at
least*. Figure 2.1 shows a *black box model* of an agent, which is embedded in its
environment. Internally, it is described by a function f which takes perception
and received messages as input and generates output in terms of performing
actions and sending messages. The mapping f itself is not directly controlled
by an external authority: the agent is *autonomous*. This very general view
fits control theoretic models of agents (see e.g., [DW91]), biological models
of agents (see e.g., [Min86] [Bro86] [Fer89]), as well as models that are based
on the knowledge-based paradigm and that define agents by their mental
states (see e.g., [BIP87] [CL90] [RG91b] [Sho93]). What makes these models
drastically differ from each other, however, is the nature of the function f
which determines an agent's behavior. This description of an agent appears

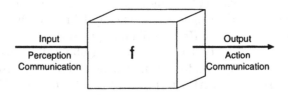

Fig. 2.1. A black box agent model

to be too general in that it names necessary preconditions for agenthood, but
not sufficient ones: for example, should a sorting algorithm that receives as
input a list and a sorting criterion and returns as output the sorted list be
looked upon as an agent? This question leads to the discussion on *legitimacy*
vs *usefulness* that goes back to John McCarthy [McC79][2]. For example, while
it is *legitimate* to consider a sorting algorithm as an agent, the key question
is whether it is really *useful* to do so. In his work on *intentionality*, Daniel
Dennett [Den87] made a very similar point in observing that the real question
is not whether a system is intentional but rather whether it can be viewed
as such *coherently*. Ascribing agenthood to machines, viewed from the perspective of the system designer, has always a strong aspect of *purpose* of the
agent; the above model is too general to capture this notion.

---

[2] This paper discusses the problem of ascribing mental qualities to machines, which
indeed is very similar to the problem of *agentification* identified by Shoham
[Sho93], i.e., ascribing the property of agenthood to machines.

This chapter provides an overview of the most influential threads in intelligent agent design. We shall start in Section 2.2 with a discussion of three related disciplines whose understanding is crucial for any work in agent design: control theory, cognitive psychology, and classical AI planning. Sections 2.3, 2.4, and 2.5 discuss three classes of intelligent agents: deliberative, reactive, and interacting agents, respectively. In Section 2.6, the concept of a hybrid architecture is discussed as a mechanism to incorporate different properties of agents that may be desirable. Finally, Section 2.7 summarizes the major issues and motivates our approach in the INTERRAP architecture (see Chapter 3).

In this chapter we focus on work, which is very closely related to ours[3]. We also leave out some areas the reader might expect to be discussed: *software agents* [GK94], *softbots* [EW94], and agent-based software engineering are only briefly discussed in Section 2.3. Work on knowledge-based distributed systems (cf. [HM90]) focuses on properties of groups of agents rather than on the individual agent, and is therefore not dealt with here. Finally, a considerable amount of research has been accomplished in the area of communication languages for agents [FF94] [MLF96] and the use of speech acts for modeling this communication [CP86] [Per90] [Lux95]. This work is partly discussed in Chapter 4; however, we place emphasis emphasis on modeling interacting agents, not on models of interaction and communication as such.

2.2 Parent Disciplines

At least three main research areas have influenced the development of intelligent agent design[4]: *Control theory, cognitive psychology,* and *classical AI planning theory.* This section provides a brief survey of these disciplines and their respective relationship to agent design.

2.2.1 Control theory

The fundamental notion in classical control theory is that of a *dynamical system* (see Figure 2.2). Such a system consists of a controller and an environment. Both the controller and the environment to be controlled are viewed as deterministic finite automata. The input of the controller is a signal which is the output of the environment; on the other hand, the input to the environment is an action provided by the controller. Classical control theory provides a mathematical model describing the interaction of these two processes.

---

[3] For a recent survey of the field, we refer to [WJ95b]. Recent collections of papers containing several other interesting approaches to agent design are [WJ95a] [WMT96].

[4] Apart from that, also the fields of object-oriented programming (see, e.g., [RBP91]) and Distributed Systems (see e.g., [Tan88]) had a fertilizing influcence on agent research.

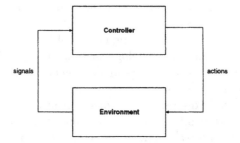

Fig. 2.2. A dynamical system

In the following, we will briefly describe the elements of a dynamical system. We introduce a set T of time points, a set X of possible states of the environment, a set U of possible inputs to the controller, and a set Y of possible outputs of the controller[5]. U corresponds to the perception of the controller, Y to the actions it may perform. Let $x(t)$, $u(t)$, $y(t)$ denote the state of the environment, the input to the controller, and the output of the controller at time $t \in T$, respectively.

The different ways a system may develop over time are represented by so-called histories of states, input, and output. These histories describe sets of mappings from time points to states. In classical control theory, the development of the state histories is restricted by a series of conditions which have to be satisfied by the environment as well as by the controller. The behavior of the environment can be classified by two restrictions. Firstly, the change of the environment as a result of actions performed by the controller is subject to certain laws. These are normally stated as difference equations of the form $x(t + 1) = f(x(t), u(t))$. Secondly, the output of the environment, i.e., the input to the controller, is directly related to the current state of the environment. This restriction is represented by a function g with $y(t) = g(x(t))$, which can be the identity function.

The task of the controller is to achieve a reaction in the environment. This can be represented by a relation $K \subset Y \times U$ describing the correct action to be executed by the controller for each possible perception. In [DW91], goals are regarded as abstract tasks that specify preferred states. Finding a sequence of actions suited to achieve a certain goal is called the *control problem*.

Very often, the control problem is split up into two sub-problems, namely the *state estimation problem* and the *regulation problem*; the former consists in determining the current state of the environment by evaluating the current

---

[5] In control theory, U is normally called output, and Y is called input; i.e., both notions are viewed from the perspective of the environment rather than from that of the controller. We use the inverse notation because we find it more convenient for a reader familiar with agent-oriented concepts.

input to the controller; the latter deals with the question of determining a suitable response given a state of the environment. This means that the controller is implemented by two functions, a *state estimator est* : $Y \mapsto X$ and a *regulator reg* : $X \mapsto U$. The output x' of *est* at time t is used as input for *reg* to compute the next action to be performed by the controller (see Figure 2.3).

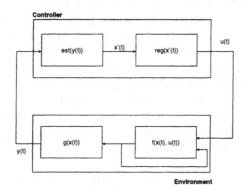

Fig. 2.3. Components of a dynamical system

Given a formal description of a dynamical system, we can now predict the behavior of the system, and, from the point of view of the controller, look at how it can achieve a desired behavior of the process to be controlled. A standard approach of classical control theory is to use integral and differential calculi in order to model both the controlling process and the process to be controlled.

Control algorithms. Having an exact model of the system to be controlled is an important precondition for building efficient controllers. Often, however, modeling the environment accurately is impossible. Two control mechanisms (i.e., mechanisms describing the interaction between the controller and the environment) have been used to deal with different settings: *feedback control* and *feedforward control*. In the case of a feedback controller, the output of the controlling process is determined from the observed behavior of the controlled process. E.g., PID[6] controllers are classical examples for feedback controllers in that they compute the control response by monitoring different aspects of the input signal as its intensity (proportional part), its accumulated value (integral part), and the way it changes (differential part). Feedback controllers are useful for applications where it is not possible to exactly predict the effects of a control action on the environment, but where the controller can closely

---

[6] Proportional–Integral–Differential.

monitor the behavior of the environment and quickly adapt its output to changes in this behavior.

In other applications, the system is not able to effectively monitor the controlled process; however, it can predict the reaction of the process to be controlled to actions performed by the controller. In these cases, the existence of an accurate model of the environment enables the controller to implement *feedforward control*.

Control theory and agent design: analogies and differences. Describing and controlling a moving body by means of differential equations seems to be a very different task from designing an intelligent agent such as a robot, a shipping company, or some interface agent assisting the employees in an office. However, at a certain level of abstraction, as argued by [DW91] and others, there is a similarity between both perspectives: in both approaches there is one process which controls and another process which is to be controlled. The controller process has an interface to the process to be controlled (the environment). It seems reasonable to associate the controller process plus its interface to the environment with an agent, and to associate the environment with the rest of the world as perceived by the agent.

The basic view of the controller as consisting of a state estimation function *est* and a regulation function *reg*, and as having the task to achieve a certain behavior of the environment, can be transformed into the AI perspective on agents and their environment in a rather straightforward manner: the agent has the ability to recognize certain classes of situations, to derive goals from these situations, and to perform actions in order to achieve these goals. In fact, the agent architecture proposed in this book incorporates *situation recognition* and *planning and scheduling* as the basic modules describing the control behavior of an agent. The difference lies in the way the function f is realized, i.e., by differential equations in the former, and by *symbolic reasoning* on an explicitly represented aspect of the world, in the latter. The function f, which models the effects of actions on the world, is implemented by the specification of postconditions for actions. The function g that specifies how an agent perceives the actual state of the world is to explain the fact that an agents' sensing of the environment may be incomplete and even erroneous, and thus needs some way of *information assessment*.

From the point of view of control, many planning approaches implement *feedforward control*: models that help to predict the reaction of the environment to actions performed by the agent are used to construct hypothetical worlds and to *plan* optimal courses of action. What has become known by the name of *reactive systems* (see Section 2.4) basically incorporates *feedback control*: the next action to be performed by the system is determined to be one of a set of situated rules whose precondition (some current input from the environment) is satisfied and which is thus able to fire.

Besides these analogies, there are of course some fundamental differences:

- Most environments are too complex to be described by differential equations: the behavior of an airport, of a shipping company, of a human enterprise, or of a manufacturing unit seem to require some sort of symbolic model.

- Classical control theory cannot cope very well with incomplete information about the environment; the explicit representation of knowledge allows an agent to reason about its knowledge and about its incompleteness; heuristics are used to deal with incomplete knowledge.

- In a multiagent environment, knowledge about the source of information is necessary in order to be able to interpret this information appropriately; in negotiating with others, an agent will have to reason about whether these other agents might be likely to tell the truth or whether they might benefit from lying. These problems cannot be adequately represented in a classical controller-environment paradigm where there is no notion of intentionality.

- Inconsistent information needs to be coped with; AI belief revision and default reasoning mechanisms offer powerful concepts for incorporating inconsistent information into a consistent world model.

To summarize, control theory is a field that has fertilized AI research to a great extent. Research on agents as it is viewed today carries on theories and concepts from control theory, trying to overcome some of its basic limits.

2.2.2 Cognitive psychology

Control theory investigates the agent-world relationship from a machine-oriented perspective. The question of how goals and intentions of a *human* agent emerge and how they finally lead to the execution of actions that change the state of the world is the subject of cognitive psychology and, in particular, of motivation theory. Motivation theory is inter alia centered around the problem of finding out why an agent performs a certain action or reveals a certain behavior. Thus, it covers the transition from motivation to action [Hec89] as illustrated in Figure 2.4. There are two subprocesses to be distin-

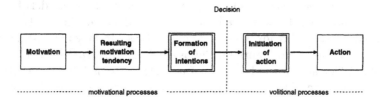

Fig. 2.4. From motivation to action

guished in the process shown in Figure 2.4 which at the same time define two basic directions in motivation theory:

- **The formation of intentions:** Starting from a set of (possibly inconsistent) motivations, the resulting motivation tendency is derived, which is the basis for the formation of intentions to act. The process of how intentions are generated from a set of latent motivation tendencies is the major research issue of motivation theory. Inter alia it has to deal with problems of selecting a consistent set of intentions from a possibly chaotic and inconsistent set of motivations.
- **Volition and action:** A second branch within motivation theory focuses on how intentions are put into practice, i.e., on how the actions of a person emerge from its intentions. Thus, having selected a set of intentions to act, at this stage a person has to decide *how and when* the actions are to be initiated.

The investigation of reasons, motivation, activation, control, and duration of human behavior goes back at least to Platon and Aristotle who defined it along three general categories: cognition, emotion, and motivation. Since then up to the 19th century, the main determinant of motivation has been seen to be situated in the human personality: a human being was looked upon as a rational creature with a free will. This approach that is best embodied by Descartes's COGITO, ERGO SUM answered the question of human motivation and decision in an intrinsic way: humans are rational, therefore they act rationally. Human behavior was regarded as being a dualist concept to the instinct-based nature of animals. An important thread of motivation theory throughout the 20th century has kept on investigating the personal determinant of motivation in the spirit of humanism (see e.g., [Ach05] [Ach10] [Mur38] [Mas43] [Mas54] [Tol52]), leading to models such as Maslow's hierarchy of human needs that has been adopted by AI researchers as a conceptual model for structuring the needs and goals of artificial agents.

In the middle of the 19th century, Charles Darwin's (1809 – 1882) theory of the origin and the behavior of species [Dar59] [Dar72] challenged the dualist perspective by postulating a strict causal determinedness of human behavior. Based on two principles of evolution, random mutation and natural selection, Darwin shifted the focus of motivation research from the person-centered to a situation-centered perspective. Darwin's theory denies the dualism between human and animal behavior and implies that many of the models for animal behaviors should also be valid for humans. Based on his work, several influential research threads have emerged, such as McDougall's instinct theory [McD08] and Freud's work on the psychological interpretation of dream contents (1900) and of the behavior of neurotic persons (1915) [Fre52]. Another consequence of Darwin's theory has been that human intelligence was viewed as a product of evolution rather than a fundamental quality which is given to humans exclusively by some higher authority. Thus, intelligence and learning became a subject of systematic and empirical research (see e.g., [Tho98] [Tho11] [Jam90] [Paw27] [Ski35]).

There are some striking parallels between the debate of reactive, behavior-based versus deliberative, plan-based systems currently held among different groups of AI researchers (see [FHN71] [Bro86] [Ste90] [Etz93]) and the discussion in motivation theory between the person-centered and situation-centered paradigms. In the case of AI, hybrid architectures (see Section 2.6) have been recently developed to combine both paradigms. In motivation theory, the insight that both situational and personal determinants influence the motivation of a human was first formulated by Kurt Lewin (1890–1947) in his theory of personality. Lewin [Lew35] described human activity as a function of both his/her personal disposition and the environment and thus prepared the way for *interactionism* that turned into one of the most influential schools of motivation theory in the 20th century. Many more recent theories, such as expectancy-value models, utilitarian decision theory, or Atkinson and Birch's Dynamic Theory of Action are influenced by Lewin's work. Since the latter theory is somehow representative for the role of psychological models for agent modeling, we shall give a brief overview.

The *dynamic theory of action* (DTA) developed in [AB70] is a model explaining the dynamics of change of motivation over time. The model especially strives after explaining change and resumption of motivation over time; special attention is paid to the transitions among different action tendencies. The model starts from a set of behavioral tendencies which can be compared to the possible goals of a person. For every point in time t and for each behavioral tendency b, the theory determines a *resultant action tendency* which determines how strong b is at time t. The tendency whose resultant action tendency is maximal is called *dominating action tendency* at time t.

The input for the DTA are an instant t in the stream of behavior, and an action tendency b which is given by a motive (personality determinant) and an incentive (situated determinant). The dynamics of the DTA is described by means of four basic forces, namely (i) an **instigating force** which is the force pushing the action tendency for b at time t; (ii) a **consummatory force** which is used to weaken the instigating force for b over time. This force is only active while the behavioral tendency b is active; (iii) An **inhibitory force** which inhibits the action tendency for b at time t; and a **force of resistance** that weakens the inhibitory force over time. The output of the DTA is the resulting tendency of action for a and t_n which is computed as a function of the four forces defined above. We refer to [Hec89] for a comprehensive presentation of the model.

Discussion. At a first glance, the DTA seems to be a well-defined formal method for deciding what goals to pursue based on information about the current situation and the mental state (motives) of a person/agent. However, the practical usefulness of the method is a controversial subject even in psychology itself due to the complexity and variety of the model parameters and of the assumptions defining the model ([Hec89, p. 475]). Looking at the DTA from the point of view of a computer scientist raises difficult questions, e.g.,

- How can motives and situations be represented and recognized?
- How can the influence of motives and situations to the basic forces F, C, I, and R be put into a computational model?
- How do the parameters have to be set to achieve a coherent change of motivation, for example to prevent continuous oscillation among different goals or behavioral tendencies?
- Can we reduce an agent to a finite set of potential behavioral tendencies?
- Assuming we can solve the problems listed so far, will such a model be tractable in practice?

However, even if we must deny the usefulness of turning the DTA directly into a computational model for intention selection based on motivations, there is a close relationship at a conceptual level between recent research work in intelligent agent design and psychological models such as the DTA.

2.2.3 Classical AI planning systems

AI planning systems such as STRIPS [FHN71] or NOAH [Sac75] view the problem-solving behavior of agents as a *sense-plan-act* cycle. A specific planning problem in STRIPS is described by an initial world state, a goal state, and a set of operators. Given such a description, the problem is to find a sequence of actions (operator executions) that turn the initial state into the goal state. Planning can thus be looked upon as search in a state space. The execution of a plan will result in some goal of the agent being achieved. In STRIPS, the agent has a symbolic representation of the world: the state of the world is described by a set of propositions that are valid in the world. The effectors of the agent are also described by a set of operators, together with a representation of the expected effects of executing the operators. These effects are represented by add- and delete lists: the add list of an operator specifies those proposition that will become valid in the world state that results from executing the operator; the delete list of an operator specifies the set of propositions that is no longer true in the resulting world state.

Some of the basic restriction made in the original AI planning framework are:

- Complete and up-to-date information.
- The world changes only by actions performed by the agent.
- Actions are deterministic, their specifications are correct; especially, actions whose preconditions are satisfied always produce the expected results and do not fail when they are executed, i.e., plan execution is not an important issue.

Subsequent work has tried to relax some of these conditions, and support e.g., the interleaving of planning and execution [McD90], replanning (e.g., the planners SIPE [Wil88] and IPEM [AIS90]), and planning under resource limitations [BD88]. In fact, much of the work in intelligent agent design, and especially in the area of deliberative approaches as they are described in Section 2.3, has been strongly influenced by the classical AI planning in preserving the symbolic representation of knowledge, skills, and goals, and in maintaining a view of the process of planning and plan execution as transitions among discrete states.

2.3 Deliberative Agents

Most agent models in AI are based on Simon and Newell's physical symbol system hypothesis [NS76] in their assumption that agents maintain an internal representation of their world, and that there is an explicit mental state which can be modified by some form of symbolic reasoning. These agents are often called *deliberative agents*. Over the past few years, an interesting research direction has explored the modeling of agents based on beliefs, desires, and intentions. Architectures following this paradigm have become known as *Belief, Desire, Intention* architectures (BDI).

2.3.1 Belief, desire, intentions

The notion of BDI (Belief, Desire, Intentions) architectures dates back to Bratman et al [Bra87] in 1987[7]; since that time, it has become a strong research area in agent design, see e.g., [RG91b] [RG92] [RG95].

The basic idea of the BDI approach is to describe the internal processing state of an agent by means of a set of *mental categories*, and to define a control architecture by which the agent rationally selects its course of action based on their representation. The mental categories are *belief, desire*, and *intentions*; in more practical BDI-approaches such as [BIP87] [RG92], these have been supplemented by the notions of *goals* and *plans*. Informally, the individual concepts are described as follows:

Beliefs of an agent express its expectations about the current state of the world and about the likelihood of a course of action achieving certain effects. Beliefs are modeled using possible-worlds semantics, where a set of possible worlds is associated with each situation, denoting the worlds the agent believes to be possible. For a detailed discussion of models and axioms of belief, we refer to [MvdHV91] [HM92] [Woo92].

Desire is an abstract notion that specifies preferences over future world states or courses of action. An important feature of desire is that an agent is allowed to have inconsistent desires, and that it does not have to believe that its desires are achievable. Thus, for example, it is possible for an

---

[7] This work had its origin in the Rational Agency project at CSLI, with the contribution of researchers such as Michael Georgeff, Kurt Konolige, Phil Cohen, Pat Hayes, and Amy Lansky.

agent to maintain a set of desires containing: (i) to be able to fly, (ii) to be rich, and (iii) not to have to work, even if the agent does not believe that it is able to fly, and even under the assumption that being rich implies having to work.

Goals: The weak definition of desires as above enforces an additional step of selecting a consistent subset of desires that an agent might pursue, i.e., goals that denote the options the agent currently has. However, there is not yet any *commitment* to specific courses of action. The notion of commitment describes the transition from goals to intentions. Additionally, it is often required that the agent believes its goals to be achievable (property of realism [CL87]). Rao and Georgeff [RG91b] require *strong realism* by putting forward the axiom that an agent's goals are a subset of its beliefs. Additional possible relationships are e.g., *weak realism* [RG91a], which means that if an agent desires a proposition, it will not believe the negation of the proposition to be inevitable, and the original notion of *realism* introduced by [CL90] (agents desire the propositions they believe in). See [RG95] for a discussion of different axiomatizations.

Intention: Since an agent is resource-bounded, it cannot pursue all its goals or options at once. Even if the set of goals is consistent, it is often necessary to select a certain goal (or a set of goals) to commit to. It is this process that is called the formation of intentions. Thus, the current intentions of an agent (its intention structure) are described by a set of selected goals together with their state of processing.

Plans: Even though they are not a conceptual ingredient of BDI theory, plans are very important for a pragmatic implementation of intentions. Bratman [BIP87] put emphasis on the fact that intentions are partial plans of actions that the agent is committed to execute in order to achieve its goals. Thus, it is possible to structure intentions into larger plans, and to define an agent's intentions as the plans it has currently adopted. In their abstract agent interpreter [RG92], Rao and Georgeff have adopted this stance. They operationalize the intentions of an agent to be the agent's current runtime stack, i.e., the current state of processing the agents plans.

In the following, we would like to give a more detailed description of two seminal approaches to BDI architectures. Bratman et al.'s agent architecture IRMA and the formalization of BDI theory by Rao and Georgeff.

2.3.2 Bratman et al.: IRMA

IRMA [BIP87] is an architecture for resource-bounded agents that describes how an agent selects its course of action based on explicit representations of its perception, beliefs, desires, and intentions. The architecture incorporates a number of modules including an intention structure, which is basically a time-ordered set of partial, tree-structured plans, a means-end reasoner, an opportunity analyzer, a filtering process, and a deliberation procedure. As soon as the agent's beliefs are updated by its perception, the opportunity analyzer is able to suggest options for action based on the agent's beliefs. Further options are suggested by the means-end reasoner from the current intentional structure of the agent. All available options run through a filtering process where they are tested for consistency with the agent's current intentional structure. Finally, options that pass the filtering process successfully are passed to a deliberation process that modifies the intention structure by adopting a new intention, i.e., by committing to a specific partial plan.

The IRMA model embodies two different views on plans: on the one hand, the plans that are stored in the plan library can be looked upon as beliefs the agent has about what actions are useful for achieving its goals. On the other hand, the set of plans the agent has currently adopted define its local intentional structure. This second view of plans as intentions has become now the most accepted paradigm in research on BDI architectures.

IRMA takes a pragmatic stance towards BDI architectures. In particular, it does not provide an explicit formal model for beliefs, goals, and intentions nor for their processing[8]. Thus, the contribution of IRMA has rather been the definition of a control framework for BDI-style agents which served as a basis to many subsequent formal refinements of BDI–concepts.

2.3.3 Rao and Georgeff: a formal BDI model

Anand Rao and Michael Georgeff [RG91c] formalized the BDI model, including the definition of the underlying logics, the description of belief , desire, and intentions as modal operators, the definition of a possible worlds semantics for these operators, and an axiomatization defining the interrelationship and properties of the BDI-operators. In contrast to most philosophical theories, Rao and Georgeff have treated intentions as first-class citizens, i.e., as a concept which has equal status to belief and desire. This allows the representation of different types of rational commitment based on different properties on the persistence of beliefs, goals, and intentions.

The world is modeled using a temporal structure with branching time future and linear past, a so-called time tree. Situations are defined as particular time points in particular worlds. Time points are transformed into one another by *events*. There are both primitive and non-primitive events; the latter are useful to model partial and hierarchical plans that are decomposable into subplans and, finally, into primitive actions. There is a distinction between *choice* and *chance*, i.e., between the ability of an agent to deliberately select its actions from a set alternatives and the uncertainty of the outcome of actions, where the determination is made by the environment rather than by the agent.

---

[8] In [Bra87], Bratman gave a theory of intentions from a philosophical point of view.

The formal language describing these structures is a variation of the Computation Tree Logic (CTL*) [ES89]. There are two types of formulae, namely *state formulae* which are evaluated at specific time points, and *path formulae* which are evaluated over a path in a time tree. There are two modal operators, *optional* and *inevitable* defined on path formulae. For a path formula ϕ, we have $optional(\phi)$ if at a particular time point in a time tree, ϕ holds of at least one emanating path; $inevitable(\phi)$ is valid if ϕ is true of all emanating paths. Additionally, the standard modal operators $\Box$ (always), $\Diamond$ (eventually), $\bigcirc$ (next), and $\bigcap$ (until) are defined for state and path formulae. Modal operators BEL, GOAL, and INTEND are introduced for beliefs, goals, and intentions, respectively.

Semantics. is defined in three parts: a semantics for state and path formulae, a semantics of events, and a semantics of beliefs, goals, and intentions. It is specified by an interpretation M that maps a standard first-order formula into a domain and into truth values, and a possible-worlds semantics for mental modalities by introducing accessibility relations for beliefs, goals, and intentions. E.g., an agent is said to believe a formula ϕ if ϕ holds true in all belief-accessible worlds. Individual worlds w are time trees represented as tuples consisting of a set of time points, a total, transitive, and backward-linear relation on the time points, and two functions $\mathcal{S}_w$ and $\mathcal{F}_w$ mapping two adjacent time points into events, representing the success and failure of events.

The semantics of events provides a mechanism for defining the success or failure of events in transforming one time point into another. In order to express success or failure of an event e, the formulae $succeeded(e)$ and $failed(e)$ are used. The semantics of these formulae is defined by using the functions $\mathcal{S}_w$ and $\mathcal{F}_w$:

$M, v, w_t \models succeeded(e)$ iff there exists t' such that $\mathcal{S}_w(t', t) = e$
$M, v, w_t \models failed(e)$ iff there exists t' such that $\mathcal{F}_w(t', t) = e$

Beliefs, goals, and intentions have a possible worlds semantics. In each situation, there are a set of belief-accessible, goal-accessible, and intention-accessible worlds which characterize those worlds the agent believes to be possible, desires to achieve, and has committed to achieve, respectively. These worlds are provided by accessibility relations $\mathcal{B}$, $\mathcal{G}$, and $\mathcal{I}$ for beliefs, goals, and intentions. Let $\mathcal{B}_t^w$ ($\mathcal{G}_t^w$, $\mathcal{I}_t^w$) denote the set of worlds which are $\mathcal{B}$ ($\mathcal{G}$, $\mathcal{I}$)-accessible from world w at time t. Let M be an interpretation with a variable assignment v. Then the semantics of beliefs, goals, and intentions is defined as:

$M, v, w_t \models \text{BEL}(\Phi)$ iff $\forall w' \in \mathcal{B}_t^w \; M, v, w_t' \models \Phi$
$M, v, w_t \models \text{GOAL}(\Phi)$ iff $\forall w' \in \mathcal{G}_t^w \; M, v, w_t' \models \Phi$
$M, v, w_t \models \text{INTEND}(\Phi)$ iff $\forall w' \in \mathcal{I}_t^w \; M, v, w_t' \models \Phi$

Axiomatization. Beliefs are axiomatized in the standard weak-S5 (KD45) modal system (for a survey of different axiomatizations for belief, we refer to [MvdHV91]). The D and K axioms are assumed for goals and intentions. That means that goals and intentions are closed under implication and that they have to be consistent. The original axiomatization provided by [RG91b] suffers from the well-known problem of logical omniscience, since, due to the necessitation rule, an agent must believe all valid formulae (see [Var86]), intends them and has the goal to achieve them.

An important requirement is belief-goal and goal-intention compatibility: For each belief-accessible world w at time t there must be a goal-accessible world that is a sub-world of w at time t. That means that a world w can only be the goal of an agent if the agent believes that it can achieve w. Goal-intention compatibility is defined analogously. From this definition, we have a subset relationship between intentions, goals, and beliefs. In [RG91b], there are additional axioms, saying e.g., that an intention will eventually lead to performing an action, or that an agent who intends a formula or has it as a goal also beliefs that it intends the formula or has them as a goal, and a corresponding set of semantic conditions which denote properties of the accessibility relations $\mathcal{B}$, $\mathcal{G}$, and $\mathcal{I}$. Additional constraints are discussed in [RG90].

Maintaining Intentions. There are three different *commitment strategies*, i.e., relationships between the current intentions of an agent and its future intentions, called *blind commitment, single-minded commitment,* and *open-minded commitment.*

- *Blind Commitment:* Once a blindly committed agent has selected an intention, it will maintain it until the agent believes that it has achieved its intentions.
- *Single-minded Commitment:* A single-minded agent will pursue an intention as long as it believes that it can still be achieved.
- *Open-minded Commitment:* Here, an agent is allowed to drop an intention not only if it no longer believes that the intention can be achieved, but also if it has this intention no longer as a goal.

Different commitment strategies are discussed in [RG91c]. Kinny and Georgeff [KG91] report a series of empirical experiments using different commitment strategies in a tileworld application. The basic results of these experiments are that the more flexible an agent's commitment strategy is, the better the agent performs in rapidly changing environments.

2.3.4 Rao and Georgeff: an interpreter for a BDI agent

A major criticism of the BDI theory as presented in [RG91b] is that the multi-modal BDI logics do not have complete axiomatizations and that no efficient implementations are available for them [RG95]; hence, so far they had little influence on the actual implementation of BDI-systems. In [RG92], the authors address this criticism by providing an abstract interpreter for a BDI agent. An abstract agent interpreter is specified that embodies the essential modules of Bratman's BDI agent (see Subsection 2.3.2). It describes the control of an agent by a processing cycle:

BDI Interpreter
initialize-state();
repeat
      options := option-generator(event-queue);
      selected-options := deliberate(options);
      update-intentions(selected-options);
      execute();
      get-new-external-events();
      drop-successful-attitudes();
      drop-impossible-attitudes();
end repeat

In each cycle, the event queue is looked up by the interpreter. A set of options is generated, i.e., goals that the agent could potentially pursue given the current state of the environment. The set of options is extended by the options that are generated by the deliberator. Finally, the intention formation step is taken in the procedure *update-intentions*. A subset of the options determined so far is selected as intentions, and the agent commits to the associated course of action. If the agent has committed to perform an executable action, the actual execution is initiated. The cycle ends by incorporating new events into the event queue, and by checking the current goals (options) and intentions whether they have been achieved, or whether they are impossible (in the case of desires) or un-realizable (in case of intentions).

This architecture provides a useful abstraction of the theoretical model, which sacrificed part of the expressional power of the theoretical framework by making certain simplifying assumptions. Using it as a basis for a practical reasoning system, however, still has problems:

- The architecture is based on a logically closed set of mental attitudes; the functions and procedures specified in the cycle (i.e., for checking successful termination or failure of goals and intentions) involve provability procedures which are not computable in general [RG95].
- The ability of the system to be *reactive* is bounded from below by the time taken to perform a cycle, since external events are taken into consideration only once in each cycle.
- Moreover, assuming no further restrictions on the option generator and the deliberation procedure, it is not clear how these can be made sufficiently fast to satisfy the real-time demands imposed on the agent.

- It should be clear that it is not possible to use a traditional theorem prover, since there is no bound on the reasoning time (and thus, on the time to act). Of course it is possible to use a theorem prover for off-line reasoning.

In order to deal with the problems of practical system design, Rao and Georgeff have proposed a number of restricting assumptions and representation choices that should result in a *practical architecture* [RG95]:

- Instead of allowing arbitrary formulae for beliefs and goals, these are restricted to be ground sets of literals with no disjunctions or implications.
- Only beliefs about the current state of the world are explicitly represented, denoting the agent's current beliefs that are prone to change over time.
- Information for means-end reasoning, i.e., about means of reaching certain world states and about the options available to the agent for proceeding towards the achievement of its goals, is represented by plans. This is achieved by representing a plan by an executable body and by a set of conditions under which a plan is an option that can be selected: an invocation condition and a precondition. The invocation condition has to be satisfied for the plan to become an option; the precondition must be satisfied for the actual execution of a plan. This representation of plans emphasizes the reactive character of plans. However, it is no help in deciding between possible options, e.g., in situations where the trigger conditions of different plans are all satisfied.
- The intentions of an agent are represented implicitly by the agent's runtime stack. Since the agent may work over a hierarchy of hierarchically related plans, it is possible that multiple stacks coexist, denoting several independent courses of action or parallel activities inside one course of action.
- In order to address the problem of ensuring reactivity within the agent cycle, it is assumed that the environment changes at rates that are comparable with the calculation cycle of the system.

The *Procedural Reasoning System* ([GL86] [GI89]) is the first implementation of a BDI architecture based on these assumptions.

In summary, it is still unclear whether it is possible to translate abstract agent specifications in multimodal logics into programs that are able to solve complex real-world problems. Rao and Georgeff's approach represents an agent's options in the context of its plans. However, it does not provide support in how the agent makes decisions among alternative options. The interpreter cycle is monolithic, and therefore does not support a special treatment of time-critical events; rather, all situations are uniformly treated by the general option generation process. Thus, it is not clear how *reactive* behavior can be realized within this paradigm.

The approach presented in this book extends the work on BDI architecture in that it is based on the assumption that different aspects of agent behavior, such as reactivity, deliberation, and interaction, correspond to and make use of different qualities of knowledge and different mechanisms for reasoning about this knowledge. This assumption is built into a layered agent architecture.

2.3.5 Shoham and Thomas: agent-oriented programming

Shoham [Sho93] proposed the framework of *agent-oriented programming* (AOP). He presented a class of agent languages that are based on a model looking upon an agent as "an entity whose state is viewed as consisting of mental components such as beliefs, capabilities, choices, and commitments". Thus, Shoham adopts the notion of an *intentional stance* as proposed by Dennett [Den87].

An AOP system is defined by three components, (i) a formal language for describing mental states of agents; (ii) a programming language with a semantics corresponding to that of a mental state; and (iii) an *agentifier*, i.e., a mechanism for turning a device into what can be called an agent, and which can thus bridge the gap between low-level machine processes and the intentional level of agent programs. In the research published so far, Shoham focused on the former two elements of AOP .

The formal language comprises the mental categories of beliefs, obligations, and capability. Obligation largely coincides with Rao and Georgeff's notion of intention and commitment. Capability is not directly represented as a mental concept in the BDI architecture; rather, it is covered by plans to achieve certain goals. Shoham's AOP language described in [Sho93] does not tackle the issue of planning and decision-making. However, Thomas' PLACA language (see below) extends AOP by planning capabilities. The language is based on a point-based propositional temporal language. The mental categories are introduced as modal operators interpreted via a possible worlds semantics (S5 system).

The programming of agents is viewed as the specification of conditions for making commitments; in [Sho93], agents commit to directly executable actions. Thomas extends the notion of commitments to plans. Shoham's programming language is an operationally-defined language that includes data structures representing both logical sentences and extra-logical operators. The control of an agent is implemented by a generic *agent interpreter* running in a two-phase loop. In each cycle, first the agent reads current messages and updates its mental state; second, it executes its current commitments, possibly resulting in further modifications of its beliefs.

AGENT0 is a simple instance of the generic interpreter. The language underlying AGENT0 comprises representations for facts, unconditional and conditional actions (both of which can be private or communicative), and commitment rules which describe conditions under which the agent will enter into new commitments based on its current mental state and on the messages received by other agents. Messages are structured according to message types;

admissible message types in AGENT0 are INFORM, REQUEST, and UNRE-QUEST. The corresponding agent interpreter instantiates the basic loop by providing functions for updating beliefs and commitments. Beliefs are updated or revised as a result of being informed or of executing an action. Commitments are updated as a result of changing beliefs or of UNREQUEST messages received by other agents.

AGENT0 is a very simple language which was not meant for building interesting applications. Important aspects of agenthood have been neglected: it does not account for motivation, i.e., for how goals of agents come into being, nor for decision-making, i.e., how the agent selects among alternative options. One basic restriction of AGENT0 was addressed by Rebecca Thomas in her 1993 Doctoral Thesis [Tho93], namely that AGENT0 agents cannot plan. The PLACA language [Tho95] is a further development of AGENT0. PLACA focuses on the agents' planning abilities and thus enables agents to communicate goal requests to other agents, whereas only requests for directly executable actions were possible in AGENT0. PLACA extends AGENT0 by introducing knowledge about goals the agent can achieve, and refines the basic agent cycle by adding a time-dependent step of plan construction and plan refinement. Thomas adopts Bratman's view of plans as intentions, i.e., the agent has a set of plans; its intentions are described by the subset of plans the agent has committed to. Whereas PLACA clearly extends the expressiveness of AGENT0 by providing the notion of plans, it does not address other restrictions such as motivation, decision-making, and the weak expressiveness of the underlying language.

2.4 Reactive Agents

In the mid-1980s, a new school of thought emerged that was strongly influenced by behaviorist psychology. Guided by researchers such as Brooks [Bro86] [Bro91], Chapman and Agre [AC87], Kaelbling [KR90], and Maes [Mae90a], architectures were developed for agents that were often called *behavior-based*, *situated*, or *reactive*. These agents make their decisions at run-time, usually based on a very limited amount of information, and simple situation-action rules. Some researchers, most notably Brooks, denied the need of any symbolic representation of the world; instead, reactive agents make decisions directly based on sensory input. The design of reactive architectures is partly guided by Simon's hypothesis [Sim81] that the complexity of the behavior of an agent can be a reflection of the complexity of the environment in which the agent is operating rather than a reflection of the agent's complex internal design. The focus of this class of system is directed towards achieving *robust* behavior instead of *correct* or *optimal* behavior.

2.4.1 Brooks: subsumption architectures

Rodney Brooks criticized that classical AI had focused too much on the issue of representation, and that abstraction was wrong in that it factors out awkward but necessary elements of intelligent behavior, such as perception and motoric skills. He advocated to "use the world as its own model", and thus to build reactive systems directly based on perception and action, which he calls the essence of intelligence. In [Bro90, p. 141], he argued that

> "[...] problem solving behavior, language, expert knowledge and application, and reason, all are pretty simple once the essence of being and reacting are available".

This observation serves as the starting point for the subsumption architectures. Whereas classical AI decomposes an intelligent system *functionally* into a set of independent information processors (e.g., vision system, planner, TMS, learning module, executor), the subsumption architecture provides an *activity-oriented* decomposition of the system into independent *activity producers* which are working in parallel, and which are directly linked to the real world by perception and action.

Brooks defined an architecture for "incrementally intelligent" agents, so-called *Creatures*. The requirements imposed on these creatures are robustness, flexibility (the ability to cope with a changing environment), the capability to maintain multiple goals, and to have a purpose in being. A system is decomposed into activity producing subsystems which may be active in parallel. Individual layers extract only these aspects of the world which are of interest to them. Thus, the representation space is cut into a set of subspaces. Between the subspaces, no representational information is passed. The lowest layers of the architecture are used to implement basic behaviors such as to avoid hitting things, or to walk around in an area. Higher layers are used to incorporate facilities such as the ability to pursue goals (e.g., looking for and grasping things while walking around).

Control is based on two general mechanisms, namely *inhibition* and *suppression*. Control is layered in that higher-level layers subsume the roles of the lower level layers when they wish to take control. Layers are able to substitute (suppress) the inputs to and to remove (inhibit) the output from lower layers for finite, pre-programmed time intervals. The ability (bias) of the robot agent to achieve its higher-level goal while still attending to its lower-level goals (e.g., the monitoring of critical situations) crucially depends on the programming of inter-layer control, making use of the two control mechanisms. Brooks was successful in building robots for room exploration, map building and route planning. However, to our knowledge, so far there are no subsumption-based robots that can do complex tasks requiring means-end reasoning and/or cooperation.

2.4.2 Agre and Chapman: Pengi

Agre and Chapman described the implementation of a theory of agent activity that is based on a view of *plans as communications* [AC87]. Whereas traditional AI planning can be characterized as *plans as programs*, i.e., as the execution of an effective procedure using an elaborated planner and a simple execution mechanism, the *plans as communications* paradigm (see also [AC90]) states that the use of plans is similar to following natural language instructions. Plans are not totally specified and, like recipes, rather suggest actions; these suggestions must be constantly interpreted by an (intelligent) execution mechanism.

The theory of activity was implemented using the Pengi system. Pengi is a program for playing the commercial video game Pengo, where a penguin is to be navigated through a two-dimensional maze populated with ice cubes and killer bees. Bees chase the penguin and, if they get too close, kill it. The penguin can kick ice cubes around; if a bee or the penguin are hit by a sliding block, they die.

Pengi consists of two subsystems: a central system and a peripheral system. The peripheral system consists of a visual system whose task is to recognize important situations (called *aspects*) by means of so-called visual routines, and of a simple motor system providing the interface to the effectoric capabilities of the agent (moving, kicking ice cubes). The central system is responsible for the agent's cognition, i.e., for recognizing and acting on important aspects of the situation. It is implemented by a combinatorial network that receives its input from the perceptual system and directs its output to the motor control system.

Pengi interacts with its environment by a set of *routines*. For example, one routine is "when you are being chased by a bee, run away". Routines are opportunistic, i.e., their activity depends on the condition part being satisfied. The underlying representational model is *indexical-functional*. That means that entities and aspects are described relative to the agent (= indexicality) instead of requiring a correspondence between symbols represented by the agent and objective individuals in the world. For example, one entity which is relevant for Pengi is *the-bee-I-am-chasing*[9]. This is indexical since the entity is described relative to the agent; it is *functional*, since entities and aspects are introduced for a specific purpose and defined in terms of the relationship to the agent. Compared to traditional logical approaches, this representation model avoids the overhead of having to instantiate variables with constants denoting objective individuals.

The model of improvised activity proposed by Agre and Chapman has some obvious limitations. For example, it is easy to specify a rule saying that whenever an agent sees a magic cube, it should kick it to get an extra score. However, it is much less obvious what a rule helping the agent to

---

[9] This example has been adopted from [AC90, p. 21].

find magic cubes as quickly as possible may look like. As all other models described so far, Pengi suffers from the inability to behave in a long-term goal-directed fashion. Moreover, it is not clear at all how the problem of arbitration between different routines that are executable at the same time can be solved in general. More recently, the authors investigated how a model based on plans as communications can be integrated into a situated theory of activity [AC90].

2.4.3 Ferber: ECO models

ECO-Problem Solving [Fer89] [FJ91] is a technique for distributed problem-solving, where the objects in the world are represented by so called ECO-agents. These actor-based agents use Agha's model of continuation [Agh86] and incorporate very simple behaviors. Problem-solving is regarded as the process of reaching a stable state in a dynamic environment. The basic parts of an ECO-system are a domain-independent set of ECO-behaviors and application-dependent domain actions and local knowledge. The local knowledge of agents can be modified by their behaviors. An agent knows about its satisfaction state, about agents on which its goal satisfaction depends, and about agents which inhibit its actions (jailers).

[FJ91] made a distinction between three general behaviors: The will to be satisfied, the will to be free, and the obligation to flee. The actions of agents are driven only by their will to be satisfied or by the obligation to flee. Interaction among agents arises by two simple mechanisms: if an agent who has the will to be satisfied cannot be satisfied, or if it is not free, it attacks jailers. In a blocks' world, this "aggression" may be implemented by sending a "go away" message to another agent. On the other hand, agents flee as an answer to aggression. The obligation to flee ensures that an agent who is told to flee will do this. Thus, the *aggression* and *flee* mechanisms provide mechanisms for conflict avoidance and conflict resolution.

The ECO model defines a simple agent model with simple mechanisms of interaction. Cooperation in this scenario emerges by local actions of the agents, and is hard to describe and to understand. In [DD92] an ECO-solution of the n-puzzle was presented where the society of ECO-tiles did a remarkably good job in finding a stable state. However, since the ECO model was developed from the point of view of Distributed Problem-Solving, goals and intentions of the agents must be coded into (rather unintuitive) dependencies and satisfaction states. In general, designing such a system is a non-trivial task.

2.4.4 Steels: behavior-based robots

In [Ste90], Steels looks upon the traditional AI approaches as not suitable for agents living in a dynamic environment. However, his approach is different in that it desists from any planning and referring to the principle of *emergent functionality* brought about by processes of self-organization which plays an important role in system theory [NP77], biology, physics and chemistry [Bab86].

The fundamental observation is that complex behavior of the system as a whole can emerge by the interaction of simple individuals with simple behavior. This describes the phenomenon of *swarm intelligence*. Steels provided the following example for self-organization in the planet scenario: robots, which have the task to collect samples located in clusters, use simple rules to indicate regions where samples may be found: if an agent carries a sample, it drops crumbs, if it carries none and detects crumbs, it picks up the crumbs again. Thus, paths are built leading to regions with high density of samples. On the other hand, agents take into account the information provided by the crumbs by following the highest concentration of crumbs. By a simulation, Steels shows that the performance of the agents can be remarkably improved by this reactive cooperation method.

As for most other reactive approaches, Steels' model suffers from the fuzziness of the underlying terms such as *self-organization*, and *emerging behavior*. The extent to which his model can be generalized, and its general usefulness as a model for intelligent agents that are able to deal with a broader range of tasks and environments, is unclear.

2.4.5 Arkin - the AuRA architecture

AuRA (Autonomous Robot Architecture) [Ark91] is an architecture for reactive robot navigation that extends Brooks' approach by incorporating different types of domain knowledge to achieve more flexibility and adaptability. Arkin made a distinction between behavioral knowledge, perceptual knowledge, and world knowledge. Behavioral knowledge is represented by a set of motor schemas, which are collections of individual motor behaviors. Examples for motor schemas include moving ahead, obstacle avoidance, staying on a path, a "noise" behavior for dealing with local minima, and mating with a docking station. The reactions of the robot to its environment are described by a *potential field* representation; the output of each motor schema is a vector representing the direction and the intended speed of the robot. Outputs of different motor schemas are combined to a resultant potential field which defines the actual movement of the robot. Perceptual schemas are strategies that provide sensory information to the motor schemas. Finally, world knowledge is used to select and reconfigure both the motor schemas and the perceptual strategies that are needed to achieve the robot's tasks.

According to this functionality, AuRA consists of five components: (i) a perception subsystem which provides perceptual input for other modules; (ii) a cartographic subsystem, i.e., a knowledge base for maintaining both a priori and acquired world knowledge; (iii) a planning subsystem which consists of a hierarchical planner and a reactive plan execution subsystem; (iv) a motor subsystem providing the effectoric interface to the actual robot; and a (v) homeostatic control subsystem which monitors internal conditions of the robot such as its energy level, and which provides this status information both to the planning subsystem and to the motor subsystem.

The coupling of planner and reacting subsystem is similar to the Procedural Reasoning System [GL86] and to Firby's RAP System (see Section 2.6). However, the approach focuses on reactivity; the underlying representation by a potential field is very application-specific and lacks generality. Moreover it is hard to see how models of other agents could be incorporated into the architecture apart from treating them as obstacles in a potential field. Thus, as in basically all reactive approaches, the cooperative abilities of AuRA robots do not exceed that of simple grouping or following behaviors (see also [Mat93] [BA95]). There is no way of expressing goals or even cooperative or synchronized plans.

2.4.6 Maes: dynamic action selection

Pattie Maes [Mae89] [Mae90b] presented a model of action selection in dynamic agents, i.e., a model the agent can use to decide what to do next. Driven by the drawbacks of both purely deliberative agents and purely situated agents, Maes argues in favor of introducing the notion of goals for situated agents. However, in contrast to traditional symbolic approaches, her model is based on the idea of describing action selection as an emergent property of a dynamics of activation and inhibition among the different actions the agent can execute. The model eschews any global control arbitrating among the different actions. An agent is described by a set of *competence modules*; these correspond to the notion of operators in classical AI planning. Each module is described by preconditions, expected effects (add and delete lists), and a level of activation. Modules are arranged in a network by different types of links: successor links, predecessor links, and conflictor links. A successor link $a \xrightarrow{s} b$ denotes that a provides the precondition for b. A predecessor link between two modules a and b is defined as $a \xrightarrow{p} b$ iff $b \xrightarrow{s} a$. Finally, a conflictor link between two modules a and b, $a \xrightarrow{c} b$ denotes that a disables b by destroying b's precondition.

Modules use these links to activate or inhibit each other in three basic ways: firstly, the *activation of successors* occurs by an executable module spreading activation forward. This method implements the concept of enforcing sequential actions. Secondly, the *activation of predecessors* provides a simple backtracking mechanism in case of a failure. Thirdly, the *inhibition of conflictors* resolves conflicts among modules. In order to avoid cyclic inhibition, only the module with the highest activation level is able to inhibit others. Activation messages increases the activation value of a module. If the activation value of a module exceeds the threshold specified by the activation level, and if its preconditions are satisfied, the module will take action.

There are two sources of new activation energy for competence modules: firstly, the current situation may increase the activation value of a module if it (partially) satisfies its preconditions. Secondly, the current goals of the agent either increase the activation value, if a goal (partly) occurs in the add list of a module, or inhibit a module, if the module's delete list undoes a protected goal.

Maes' approach extends purely reactive approaches by introducing the useful abstraction of goals. However, there are problems: firstly, the notion of *emerging* of the action selection process makes the resulting behavior of an agent hard to understand, even harder to predict, and almost impossible to verify, i.e., it is a difficult task to design a system with defined functionality. The activation level itself does not have a clear semantics apart from the fact that a higher level dominates a lower one. How well an agent performs crucially depends on finding good settings of the activation values, which not only depends on characteristics of the problem domain, but also on the size of the module network [Mae89, p. 69].

The hybrid approaches for designing agents that are discussed in Section 2.6 try to avoid some of these drawbacks by explicitly defining and arranging different functionalities in different layers that interact through a well-defined interface.

2.5 Interacting Agents

Distributed Artificial Intelligence (DAI)[10] deals with coordination and co-operation among distributed intelligent agents. So far, its focus has been on the coordination process itself and on mechanisms for cooperation among autonomous agents rather than on the structure of these agents. In Subsection 2.5.1, we will give a brief overview of agent interaction and related work from the perspective of agent architectures.

2.5.1 Main topics in agent interaction

Communication. Communication naturally plays a significant role in the interaction among agents. Since the beginning of the 1990s, there has been a considerable research effort in setting up standardized communication languages. The Knowledge Query and Manipulation Language (KQML) [FF94] that has been developed as a part of the ARPA Knowledge Sharing Effort is a good example for research dealing with developing languages and protocols

---

[10] See [BG88] [GH89] for collections of papers that provide a good overview of DAI research by the end of the 1980s. More recent work on DAI can be found in the annual proceedings of the Distributed AI Workshop (until 1994) and in the proceedings of the International Conference on Multiagent Systems (ICMAS) (biannually since 1995).

for exchanging information and knowledge. The language model of KQML is divided into three layers: The *communication layer* describes the lower-level communication parameters. The *message layer* which forms the core of the language identifies the underlying protocol and supplies a performative which is attached to the message content by the sender a, and the *content layer*. Examples for performatives are *assertions*, *query*, and *command*. Finally, the *content layer* contains the actual contents of the message in an agreed-upon language, e.g., KIF [GeF92].

Like many other approaches for the description of agent communication (see [Lux95] for an overview), the application-oriented work done in KQML has its roots in *speech act theory* [Aus62] [Sea69]. The idea underlying this theory is to treat communication as a specific type of action, which enables the use of speech acts in planning [CP86] [Per90] [LS95]: communicative acts such as the performatives in KQML can be provided with pre- and postconditions that specify the expected effects of sending a message to the mental state of the recipient.

Game theory and agent interaction. Rosenschein and Zlotkin contributed a game theoretic analysis of interaction among rational agents (see e.g., [Ros85] [ZR93] [RZ94]). For each individual agent, a decision matrix represents a preference distribution over alternatives that can be processed by the tools of decision and game theory in order to derive rational decisions given the existence of other individual rational agents.

Distributed Problem-Solving (DPS). Another important thread within DAI deals with the performance in a given task by using a set of distributed problem solvers[11]. The focus of this work is on mechanisms for task decomposition, on protocols for the allocation of tasks to problem solvers (such as the well-known Contract Net Protocol [DS83] and various of its extensions, e.g., [FMP95b]), on coordination among the problem solvers, and on the synthesis of a global solution from the partial solutions returned by the individual problem solvers. Thus, the distributed problem solving approach assumes the existence of intelligent agents, but does not contribute to their description.

Multiagent Planning. An area closely related to distributed problem solving is multiagent planning, i.e., the generation and execution of plans *for* and/or *by* multiple agents. Georgeff [Geo83] and Rosenschein [KR89] have investigated planning for multiple agents from a perspective of distributed problem solving. Von Martial [Mar90] [KvM91] has investigated relationships among plans of multiple agents. Grosz and Sidner [GS90] have described the use of *shared plans* in discourse. More recent approaches that focus on planning *by* multiple agents are [KLR+92], [ER93b] [ER93a]. Again, this work focuses on coordination mechanisms rather than on the description of agents.

---

[11] See e.g., [CDL87] [DDL89] [DLC89] [DL94] for examples for this research direction.

Conflicts, Cooperation, and Negotiation. A considerable amount of research work has been spent on the exploration of conflict resolution and cooperation based on negotiation. Most of this work stands firmly in the DPS tradition. For example, [CKLM91] describe a multistage negotiation protocol for distributed planning and constraint satisfaction processes. Sycara has explored negotiation mechanisms based on argumentation and persuasion to resolve conflicts among non-cooperative agents [Syc87] [KNS93]. Mark Klein [Kle91] has described a methodology for the resolution of conflicts in concurrent engineering and design applications.

Summary. In summary, research on coordination among agents has focused on mechanisms and coordination methodologies and, to a great extent, neglected aspects of describing the individual agent. However, some recent work deals with the incorporation of cooperative abilities into an agent framework. In the following, four approaches that are related with the work presented in this book shall be described in more detail: Fischer's MAGSY, the GRATE* architecture by Jennings, Steiner's MECCA system, and the COSY architecture developed by Sundermeyer and Burmeister.

2.5.2 Fischer: MAGSY

Klaus Fischer developed the MAGSY [Fis93b] system, a language for the design of multiagent systems. The agent model underlying MAGSY is fairly simple. A MAGSY agent consists of a set of facts representing its local knowledge, a set of rules representing its strategies and behavior, and a set of services that define the agent's interface. MAGSY was implemented by extending the OPS-V language [For82] by a set of communication services. Thus, facts and rules are basically OPS-V facts, and rules. An agent can request a service offered by another agent by communication.

Fischer demonstrates the applicability of MAGSY by the application of decentralized cooperative planning in an automated manufacturing environment. Agents are e.g., robots, and different types of machines like heating cells or welding machines. The domain plans of the robots are represented as Petri nets, which are translated into a set of rules, so-called *behaviors*. These behaviors are procedures that interleave planning with execution.

The MAGSY language enables the efficient and convenient implementation of multiagent systems. It provides a variety of useful services and protocols to establish multiagent communication links. Clearly, MAGSY inherits both the positive and the negative properties of rule-based programming languages: on the one hand, there is concurrency and the suitability for modeling reactive agents; on the other, there is the flat knowledge representation and the awkward way to represent sequential programs. Cooperation between agents is hard-wired by connections between the Petri nets representing behaviors of different agents. Thus, MAGSY does not support reasoning about cooperation.

2.5.3 Jennings: GRATE*

The focus of Jennings' work on the GRATE* architecture [Jen92b] was on cooperation among possibly preexisting and independent intelligent systems, through an additional *cooperation knowledge layer*. The problem solving capability of agents were extended by sharing information and tasks among each other. GRATE* is an architecture for the design of interacting problem solvers. A general description of cooperative agent behavior is represented by built-in knowledge. Domain-dependent information about other agents is stored in specific data structures (*agent models*).

GRATE* consists of two layers, a *cooperation and control layer* and a *domain level system*. The latter can be preexisting or purpose built; it provides the necessary domain functionality of the individual problem solver. The former layer is a meta-controller operating on the domain level system in order to ensure that its activities are coordinated with those of others in the multiagent system [Jen92a]. Agent models hold different types of knowledge: the *acquaintance model* includes knowledge the agent has about other agents; the *self model* comprises an abstracted perspective of the local domain level system, i.e., of the agent's skills and capabilities. The cooperation and control layer consists of three submodules, representing the interplay among local and cooperative behavior: The *control module* is responsible for the planning, execution, and monitoring of local tasks. The *cooperation module* handles processes of cooperation and coordination with other agents. The *situation assessment module* forms the interface between local and social control mechanisms. It is thus responsible for the decision to choose local or coordinated methods of problem solving.

Clearly, Jennings' focus was on the cooperation process. However, he went beyond work discussed before by defining a two-layer architecture that embeds cooperation into a domain level system. The architecture does not address more subtle questions of agent behavior, such as how to reconcile reactivity and deliberation. Rather, these problems are expected to be solved within the domain level system.

2.5.4 Steiner et al: MECCA

In the MECCA architecture [SBKL93] [Lux95] [LS95], an agent is regarded as having an application-dependent *body*, a *head* whose purpose is to actually agentify the underlying system, and a *communicator* which establishes physical communication links to other agents. This view supports the construction of multiagent systems from second principles. Agent modeling is addressed in the design of the agent's head. It is described by a *basic agent loop* consisting of four parallel processes: *goal activation*, *planning*, *scheduling*, and *execution*. In the goal activation process, relevant situations (e.g., user input) are recognized and goals are created that are input to the planning process. There, a partially ordered plan structure is generated corresponding to a set of possible courses of action the agent is allowed to take. The scheduler instantiates (serializes) this partially ordered event structure by assigning time points to actions. The execution of actions is initiated and monitored by the execution process.

All control processes in the basic loop may involve coordination with other agents, leading to joint goals, plans, and commitments, and to the synchronized execution of plans. Cooperation is based on speech act theory: MECCA provides a set of cooperation primitives (e.g., INFORM, PROPOSE, ACCEPT) which are treated by the planner as actions, i.e., whose semantics can be described by preconditions and effects. This allows the planner to reason about communication with other agents as a means of achieving goals [LS95]. Moreover, cooperation primitives are the basic building blocks of communication protocols, so-called cooperation methods.

MECCA provides an architecture for deliberative interacting agents having a symbolic representation of the world. Questions of reactivity were not addressed by the originally proposed architecture, since it was not especially important for the domains of application that were pursued at that time. The current version of the system includes a *daemon* mechanism which provides restricted possibilities to shortcut the agent loop in specific contingency situations and to initiate a kind of procedure. However, the semantics of this mechanism is not clearly specified.

2.5.5 Sundermeyer et al: COSY

The COSY agent architecture [BS92] describes an agent by *behaviors*, *resources*, and *intentions*. The behavior of an agent is classified into perceptual, cognitive, communicative, and effectoric, each of which is simulated by a specific component in the COSY architecture. Resources include cognitive resources such as knowledge and belief, communication resources such as low-level protocols and communication hardware, and physical resources. e.g., the gripper of a robot. Intentions are used in a sense that differs from [CL90] [RG91b]: there are strategic intentions modeling an agent's long-term goals, preferences, roles and responsibilities, and tactical intentions that are directly tied to actions, representing an agent's commitment to his chosen course of action.

The individual modules of COSY are ACTUATORS, SENSORS, COMMUNICATION, MOTIVATIONS, and COGNITION. The former three are domain-specific modules with their intuitive functionality. The motivations module implements the strategic intentions of an agent. The cognition module evaluates the current situation and selects, executes, and monitors actions of the agent in that situation. It consists of four subcomponents, a *Knowledge Base, Script Execution Component, Protocol Execution Component*, and *Reasoning and Deciding Component*. The application specific problem solving knowledge is encoded into plans. There are two types of plans stored in a plan library: *scripts* describing stereotypical courses of action to achieve certain goals, and *cooperation protocols* describing patterns of communication [BHS93]. Scripts are monitored and executed by the Script Execution Component, handing over the execution of primitive behaviors to the actuators, and protocols to the Protocol Execution Component. The Reasoning and Deciding Component is a general control mechanism, monitoring and administering the reasoning and decisions concerning task selection and plan selection, including the reasoning and decisions concerning intra-script and intra-protocol branches. In [Had95], Haddadi has provided a deeper theoretical model by extending Rao and Georgeff's BDI model by a theory of commitments and by defining mechanisms allowing agents to reason about how to exploit potentials for cooperation by communicating with each other. However, there still is a wide and partially not intuitive gap between Haddadi's theory and the actual implementation in the COSY model.

Similar to the MECCA architecture, COSY focuses on aspects of cooperation among rational agents. It offers a simple planning mechanism based on planning from second principles. The basic idea to use perception and intentions to guide the decisions of an agent is very similar to the ideas pursued in BDI architectures. However, recent implementations of COSY only admit very simple selection strategies and make very restrictive assumptions, e.g., a total priority ordering on the possible strategic intentions an agent can pursue. Moreover, there is no explicit concept of reactivity supported by the architecture, although the general structure of the cognition module would not be a hindrance to the design of such a mechanism.

2.6 Hybrid Architectures

The approaches discussed so far suffer from different shortcomings: whereas purely reactive systems have a limited scope insofar as they can hardly implement goal-directed behavior, most deliberative systems are based on general-purpose reasoning mechanisms which are not tractable, and which are much less reactive. One way to overcome these limitations in practice, which has become popular over the past few years (see [Bro86] [Kae90] [BM91] [Fer92] [Fir92] [LH92] [Dab93] [BKMS96] [SP96]), are layered architectures. Layering is a powerful means for structuring functionalities and control, and thus is a valuable tool for system design supporting several desired properties such as reactivity, deliberation, cooperation, and adaptability. The main idea is to structure the functionalities of an agent into two or more hierarchically organized layers that interact with each other in order to achieve coherent behavior of the agent as a whole. Layering offers the following advantages:

- It allows to modularize an agent; different functionalities are clearly separated and linked by well-defined interfaces.
- This makes the design of agents more compact, increases robustness, and facilitates debugging.

- Since different layers may run in parallel, the agent's computational capability can be increased in principle by a linear factor[12].
- Especially, the agent's reactivity can be increased: while planning, a reactive layer can still monitor the world for contingency situations.
- Since different types and partitions of knowledge are required for the implementation of different functionalities, it is often possible to restrict the amount of knowledge an individual layer needs to consider in order to make its decisions. For example, a reactive layer should only use information about the current state of the world, whereas a planning layer has to make use of a history as well as of information about goals, commitments, and available plans. Again, this restriction allows the agent to act effectively and helps increasing practical reactivity of the agent.

These advantages have made layering a popular technique that has been mostly used to reconcile reaction and deliberation. In the following, three layered approaches are presented: architectures based on Firby's RAPs, the planner-reactor architecture proposed by Lyons and Hendriks, and Ferguson's Touring Machines architecture.

2.6.1 Firby, Gat, Bonasso: reactive action packages

Firby's work [Fir89] [Fir92] has been most influential in research on integrating reaction and deliberation in the area of AI planning and robotics. In this paragraph, we outline Firby's original work and two of its recent extensions.

The RAPs System. Firby focuses on the interface between continuous and symbolic robot control, i.e., to tackle the problem how to turn symbolic actions into continuous processes that the robot physically executes and to use task-specific sensing routines to support these processes. He argues that having symbolic actions allows the designer of a robot to

> preserve the convenient fiction of "primitive actions" for use in planning without requiring that they all be programmed into the control system in advance. [Fir92, p. 62]

The RAPs (Reactive Action Packages) system describes an integrated architecture for planning and control. The underlying agent architecture consists of three modules, a *planning layer*, the *RAP executor*, and a *controller*. The planning layer produces sketchy plans for achieving goals using a world model representation and a plan library. The RAP executor fills in the details of the sketchy plans at run-time. The expansion of vague plan steps into more detailed instructions (methods) at run-time reduces the amount of planning uncertainty and thus largely simplifies planning. If incorrect methods are

---

[12] It is obvious that a part of this gain will be used up by an increased need for coordination among the layers.

selected at run-time, the RAP executor is able to recognize failure[13] and to select alternative methods to achieve the goal. Apart from providing the ability of controlling the process of achieving goals in a reactive manner, and thus from providing the interface between subsymbolic continuous and symbolic discrete representation and reasoning, the RAP executor provides a set of abstract *primitive actions* to the planner. We have adopted this view in the INTERRAP architecture.

The controller provides two kinds of routines that can be activated by requests from the RAP executor and that deliver results to that module: active sensing routines and behavior control routines. Sensing routines are useful to provide lacking information about the current world state. Behavior routines are continuous control processes that change the state of the physical environment. Examples for behavior routines are collision avoidance, visual tracking, or moving to a specified direction.

In [Fir94], the control of continuous processes (i.e., the interplay of the RAP executor and the controller) is elaborated by describing an extension to the RAPs representation language and the semantics for task nets. That paper makes the idea of modeling the behavior of a robot by a set of interacting processes acting on input from the environment more explicit; it addresses the problems of synchronizing the expansion of plans with events in the world and allows to represent multiple, non-deterministic outcome of operations. In this context, Firby argues against using the notions of success and failure to describe the outcome of these processes, since they represent a very strongly restricted (and only little helpful) interpretation of possible outcomes of task execution. Introducing a broader spectrum of status signals that may be sent to the executor by the controller allows more reasonable treatment of various exceptions by the higher system layers.

ATLANTIS. Gat [Gat91b] [Gat92] describes the heterogeneous, asynchronous architecture ATLANTIS that combines a traditional AI planner with a reactive control mechanism for robot navigation applications. ATLANTIS consists of three control components: a *controller*, a *sequencer*, and a *deliberator*. The controller is responsible for executing and monitoring the primitive *activities*[14] of the agent. [Gat91a] defines a language for modeling the often nonlinear and discontinuous control processes. The controller thus connects to the physical sensors and actuators of the system. The deliberator process performs deliberative computations which may be time-consuming, such as planning or world modeling. Between the two components stands the sequencer which initiates and terminates primitive activities by activating

---

[13] The underlying assumption is that of a *cognizant failure*: it is not required that no failure occurs, but that virtually all possible failures may be detected if they occur, and that repair methods can be applied to recognized failures.

[14] Note that Gat's notion of activities is different to the notion of *action* as it is used in classic AI in that activities describe continuous time-consuming processes that are initiated and terminated by *decisions* rather than atomic discrete state transitions.

and deactivating control processes in the controller, and which maintains the allocation of computational resources to the deliberator, by initiating and terminating deliberation with respect to a specific task. As in the RAPs system, the sequencer maintains a task queue; each task described by a set of methods together with conditions for their applicability. Methods describe either primitive activities or subtasks; in the former case, the corresponding module in the controller is activated; the latter case is handled by recursive expansion. ATLANTIS extends Firby's original work by allowing to control activities instead of primitive actions, and provides a bottom-up flow of control: in RAPs, tasks are installed by the planner whereas they are initiated in the sequencer in Gat's architecture.

The $3T$ architecture. In [BKMS96], Peter Bonasso and colleagues have defined the layered architecture $3T$ which enhances the RAPs system by a planner. In particular, $3T$ consists of three control layers: a reactive skill layer, a sequencing layer, and a deliberation layer. The reactive skill layer provides a set of situated skills. Skills are capabilities that, if placed in the proper context, achieve or maintain particular states in the world. The sequencing layer is based on the RAPs system. It maintains routine tasks that the agent has to accomplish. The sequencing layer triggers continuous control processes by activating and deactivating reactive skills. Finally, the deliberation layer provides a deliberative planning capability which selects appropriate RAPs to achieve complex tasks[15]. This selection process may involve reasoning about goals, resources, and timing constraints. Using $3T$, different robot systems have been implemented, such as a robotic wheelchair, diverse manipulation tasks, and robot errand running.

Compared to Gat's work, $3T$ uses a more powerful planning mechanism; moreover, reactivity, i.e., the ability to react to time-critical events, is implemented at the skill layer in $3T$, whereas it is partly a task of the sequencing layer in ATLANTIS. Both of these extensions make the sequencing layer in $3T$ more compact and easier to handle.

Discussion. RAPS, ATLANTIS, and $3T$ are powerful systems supporting the integration of deliberation and reaction for robotics applications. However, they do not address issues of motivation (how do goals emerge?), decision-making, and goal selection (If there are several alternatives to achieve a goal, which should be chosen?). While $3T$ can cope with some of these questions, it assumes a world in which there is only one robot. However, this restriction should be clearly differentiated against the single-agent assumption that is inherent to classical AI planning systems such as STRIPS [FHN71]: RAPs, ATLANTIS, and $3T$ *do* account for uncertainty and failure of actions due to the possible existence of other agents. The point is that they do not support making use of the existence of other agents in that they provide neither communication facilities nor the ability to reason about other agents beliefs,

---

[15] $3T$ uses the AP system (Adversarial Planner) [ES94].

goals, or plans, nor the possibility of performing tasks cooperatively. Any such reasoning has to be hard-wired into the planning, sequencing, and execution layers; it is not supported by the systems as such.

2.6.2 Lyons and Hendriks: planning and reaction

In [LH92] a practical approach towards integrating reaction and deliberation in a robotics domain is introduced based on the planner-reactor model proposed by Bresina and Drummond [DB90] in their ERE architecture incorporating planning, scheduling, and control[16]. Whereas Drummond and Bresina's model focused on the *anytime* character of the architecture, Lyons and Hendriks put emphasis on the task of producing timely, relevant actions, i.e., on the task of qualitatively reasonable behavior.

The basic structure of Lyons and Hendriks' architecture is described by a planner, a reactor, and a world (which is, in the spirits of control theory, cf. Subsection 2.2.1, looked upon as a part of the system to be described). In contrast to the hybrid approaches discussed so far, in Lyons and Hendriks' model planning is looked upon as incrementally adapting the reactive system which is running concurrently in a separated process by bringing it into accordance with a set of goals. Thus, the planner can iteratively improve the behavior of the reactive component.

The reactor itself consists of a network of *reactions*, i.e., sensory processes that are coupled with action processes in a sense that the sensory processes initiate their corresponding action processes in case they meet their trigger conditions. It can act at any time independently from the planner, and it acts in real time. The planner can reason about a model of the environment (EM), a description of the reactor (R), and a description of goals (G) that are currently to be achieved by the reactor as well as constraints imposed by these goals. The task of the planner is to continuously monitor whether the behavior of R conforms to G. If this is not the case, the planner incrementally changes the configuration of R by specifying *adaptations*. On the other hand, the reactor can send collected sensory data to the planner allowing the latter to predict the future state of the environment. Adaptations of the reactor include removing reactions from the reactor and adding new reactions.

Similar to Firby's and Gat's approaches, Lyons and Hendriks' model is restricted to the single-agent case and does not support the modeling of agents in a multiagent world. The model of adaptation of reactors leaves open some questions, e.g., that of on-line adaptation: how are inconsistencies dealt with that result from modifying a running system? Moreover, it does not address the question of general situation recognition mechanisms enabling the planner to decide whether the behavior of the reactor conforms with the goal specification and its constraints or not.

---

[16] A similar paradigm has been proposed by McDermott [McD91].

2.6.3 Ferguson: Touring Machines

In [Fer92], Ferguson describes a layered control architecture for autonomous, mobile agents performing constrained navigation tasks in a dynamic environment. The *Touring Machines* architecture consists of three layers, a *reactive layer*, a *planning layer*, and a *modeling layer*. These layers operate concurrently; each of them is connected to the agent's sensory subsystem from which it receives perceptual information, and to the action subsystem to which it sends action commands. The reactive layer is designed to compute hard-wired domain-specific action responses to specific environmental stimuli; thus, it brings about reactive behavior. On the other hand, the planning layer is responsible for generating and executing plans for the achievement of the longer-term relocation tasks the agent has to perform. Plans are stored as hierarchical partial plans in a plan library; based on a topological map, single-agent linear plans of action are computed by the agent. Planning and execution are interleaved to allow to cope with certain forms of expected failure. The modeling layer provides the agent's capability of modifying plans based on changes in its environment that cannot be dealt with by the replanning mechanisms provided by the planning layer. In addition, the modeling layer provides a framework for modeling the agent's environment, and especially, for building and maintaining mental and causal models of other agents.

The individual layers are mediated by a control framework that coordinates their access both to sensory input and to action output. This is described by means of a set of context-activated control rules. There are two types of rules: *censors* and *suppressors*[17]. Censors filter selected information from the input to the control layers; suppressors filter selected information (i.e., action commands) from the output of the control layers. Thus, undesirable interactions between layers may be prohibited by disabling the activation of this layer through denying access to certain information that would predictably trigger activation, or by forbidding the layer to act.

Ferguson's solution of the control problem has certain problems. The unrestricted concurrent access of the control layers to information and action and the global (i.e., for the agent as a whole) control rules imply a high design effort to analyze, predict, and prevent possible interactions among the layer. Since each layer may interact with any other layer in various ways either by being activated through similar patterns of perception or by triggering contradictory or incompatible actions, a large number of control rules are necessary. Thus, for complex applications, the design of consistent control rules itself is a very hard problem. The architecture presented in this book provides an architectural solution to simplify this problem.

As regards multiagent aspects, Ferguson's model is certainly the most advanced of the approaches discussed in this section in that it includes the

---

[17] Ferguson has adopted this notion from Minsky's Society of Mind [Min86]. It is also similar to the suppression/inhibition mechanism proposed by Brooks (cf. Section 2.4).

modeling of other agents in the modeling layer. However, the knowledge obtained this way is merely used to guide the agents local decisions; thus, the *Touring Machines* architecture does not give support to cooperative planning and problem solving based on communication.

2.7 Bottom Line

So far, we have described architectures for reactive agents, for deliberative agents, and for interacting agents. Hybrid architectures have been presented that integrate reaction and deliberation. Figure 2.5 illustrates a map of intelligent agent design. It presents the area of agent design as a part of Artificial Intelligence, which itself has been strongly influenced by areas such as control theory and cybernetics (discussed in Section 2.2.1), cognitive psychology (Section 2.2.2), and economics and sociology[18]. The vertical ordering of the individual approaches is approximately chronological. Based on the AI sub-

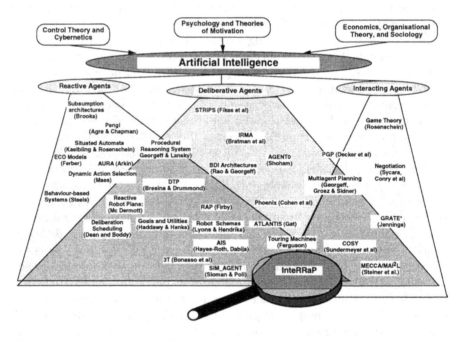

Fig. 2.5. Intelligent agent design: an area map

---

[18] This field plays a role especially in the part of intelligent agent design related with distributed AI. It has partly been discussed in Section 2.5. For more comprehensive discussions, we refer to [Axe84] [Gas91].

areas of *reactive systems, AI planning and knowledge representation*, and *Distributed AI*, three directions of intelligent agent design have been developed over the past decade:

- Reactive architectures (see Section 2.4).
- Deliberative architectures (see Section 2.3).
- Architectures for interacting or "social" agents (see Section 2.5).

A considerable amount of research effort has been dedicated to the exploration of how reaction and deliberation can be reconciled: hybrid architectures, algorithmic solutions (reactive and anytime planning [BD88] [DB90] [McD91] [RZ91] [BD94]), and approaches integrating procedural knowledge (e.g., the Procedural Reasoning System described in [GL86] [GI89]). On the other hand, there are virtually no approaches explicitly supporting the ability of agents to interact and to cooperate with other agents within a general agent architecture. It is a main contribution of this book to provide an architecture filling that gap: the INTERRAP model presented in the following supports the modeling of reactive, goal-directed, and interacting agents by providing (1) a set of hierarchical control layers, (2) a knowledge base that supports the representation of different abstraction levels of knowledge, and (3) a well-defined control structure that ensures coherent interaction among the control layers. Thus, INTERRAP extends the power of existing architectures that support a reconciliation of reaction and deliberation by explicit models of agent interaction.

3. The Agent Architecture INTERRAP

3.1 Introduction

Autonomous agents that perform tasks in a dynamic multiagent environment have to fulfill real time and real world requirements. These have to be built into their underlying design architecture. Throughout this work, we shall focus on four of these requirements:

- **Situated behavior:** Agents must recognize unexpected events and react timely and appropriately to them.
- **Goal-directed behavior:** Agents must select their actions based on the ends they want to achieve and taking into account their available means.
- **Efficiency:** Tasks are to be solved *efficiently*; often, real-time constraints have to be satisfied. This can be accomplished by providing access to a set of "hard-wired" procedures with guaranteed execution properties.
- **Coordination:** Agents must cope with the presence of other agents and with the positive and negative consequences arising from interactions.

In Chapter 2, we have seen that previous approaches have mainly concentrated either on the combination of reactivity and deliberation on the one hand, or on the modeling of agent interaction and cooperation on the other. So far, little research has been dedicated to integrating the above requirements into one agent architecture. This book aims to fill this gap. The principal idea underlying the INTERRAP model presented in this chapter is to define an agent by three interacting control and knowledge layers:

- A *behavior-based layer* incorporating reactivity and procedural knowledge for routine tasks.
- A *local planning layer* that provides the facilities for means–ends reasoning for the achievement of local tasks and to produce goal-directed behavior.
- A *cooperative planning layer* that enables agents to reason about other agents and that supports coordinated action with other agents.

These layers and the control architecture defined for them combine reactive and deliberative reasoning, and incorporate the ability to interact with other agents.

Layering is a powerful technique that allows to integrate different functionalities or levels of abstraction. In i, this technique is used in two ways:

firstly, it reconciles different functionalities that compete for the computational resources of a resource-bounded agent: reactivity, local deliberation, and reasoning about other agents. Secondly, it provides different levels of representational abstraction: object-level knowledge, local meta-knowledge, and cooperative knowledge. By restricting the access of different functional layers to different portions of knowledge, this classification restricts the representational complexity of the lower functional layers, while providing the higher functional layers with sufficient information to fulfill their purpose.

INTERRAP is a BDI architecture: the informational, motivational, and deliberative state of an agent [RG95] is described by means of beliefs, goals, a somewhat generalized version of plans, and intentions. Agent input (perception) is linked to agent output (action) by a set of functions that explain the interrelationship between the mental categories of an agent. However, in contrast to BDI architectures like [BIP88] [RG91b] [BS92], we distribute the mental categories as well as the functional relationships between them over three hierarchical layers and connect them by a hierarchical control mechanism. Our objective in doing so is to find a pragmatic solution for the problem of reconciling the different requirements stated above.

The structure of this chapter is as follows: In Section 3.2, we give an informal characterization of the conceptual agent model underlying INTERRAP. Section 3.3 defines the INTERRAP control architecture with the individual modules and their interaction. Then, the individual control layers of INTERRAP are explained in detail in Sections 3.4 to 3.6.

3.2 A Conceptual Agent Model

In Section 2.1, we have shown a reductionist model of an agent given by a functional black box, receiving perception as input and producing a certain action as output. In this section, we shall paint a more complete picture of this box. We present the conceptual agent model underlying INTERRAP, which is that of a layered BDI architecture.

3.2.1 Overview

In [RG95], Rao and Georgeff classified the mental state of an agent in an informational, a motivational, and a deliberative state. That way, they obtained three categories that map naturally into the basic building blocks of the BDI approach: beliefs, desires, and intentions. While we keep to their model as regards these basic elements, the focus of this book is not to provide a new formal model for beliefs, desires, and intentions. We are interested in the dynamic control processes of an agent rather than in the static description. Therefore, the reader will look in vain for a modal logic description of beliefs, goals, and intentions; neither will we provide an axiom system that formally defines the relationship between different mental attitudes based on a possible worlds model. Instead, the conceptual model that we describe in this section is informal; it is then mapped into a pragmatic operational control model in the following sections.

Figure 3.1 shows the conceptual model of an INTERRAP agent. The model has been influenced by the abstract agent architecture proposed by Bratman et al. in [BIP87]. It describes both the ingredients of the mental state of an INTERRAP agent and the functional connection of these ingredients defining the input into and the output of an agent. Both topics are discussed in more detail in the following paragraphs.

3.2.2 Mental categories

Figure 3.1 assembles the mental state of an agent from different components:

- The current *perception* of the agent.
- A set of *beliefs* denoting its informational state.
- A set of *situations* that describe relevant structured portions of the agent's beliefs.
- A set of (context-independent) *goals* that the agent might possibly have.
- A set of *options* representing the agent's motivational state. Based on the current situation, a set of context-dependent *options* is selected from the set of possible goals. Thus, options are goals the agent might pursue given certain environmental conditions[1].
- A set of *intentions* defining the deliberative state of the agent, i.e., the options to which the agent has committed itself, and defining the next *action* that the agent will perform.
- A set of operational primitives linking the motivational state of an agent to its deliberative state.

In contrast to Bratman's original architecture, however, the components of the mental state of an INTERRAP agent, with the exception of perception, are layered. *Beliefs* are split into a *world model*, a *mental model*, and a *social model*. The world model contains object-level beliefs about the environment; the mental model holds meta-level beliefs the agent has about itself; the social model holds (meta-level) beliefs about other agents.

While the concept of beliefs is well-suited to *describe* the relationship between the informational state and the motivational state of an agent, it is too unstructured to be very useful for its *computation*. The initiation of action is always triggered by specific *situations*, i.e., relevant subsets of the

---

[1] As the difference between goals and options is not representational but operational, we shall use the more popular notion of goals also to denote options; often, we shall distinguish between the two by referring to an agent's *possible goals* and *current goals*, accordingly. The reader should keep in mind that the latter notion corresponds to options, and that a selection process is required to derive these from the possible goals of an agent.

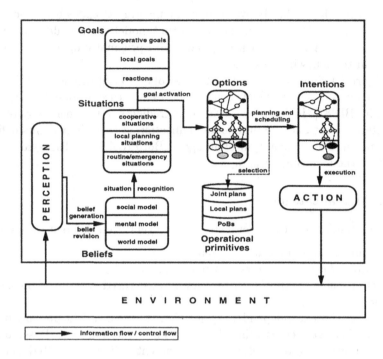

Fig. 3.1. The conceptual agent model

agent's beliefs. *Situations* are a useful representational abstraction of classes of world states that are of interest for an agent. In accordance with the three classes of beliefs, we draw a distinction between three classes of situations:

- behavioral situations, which are a subset of the agent's world model;
- situations requiring local planning, the description of which is based both on the world model and on the mental model;
- situations requiring cooperative planning, the description of which additionally contains parts of the agent's social model.

Goals are classified in reaction goals, local goals, and cooperative goals. Local goals correspond to the standard notion of a goal in BDI architectures (see Section 2.3.1); cooperative goals are shared among a group of agents. Reaction goals differ from the common notion of goals: they denote short-term goals that are triggered by external events and that require a fast reaction or the execution of a routine procedure.

Operational primitives (OPs) enable the agent to do means-ends reasoning about how to achieve certain goals. OPs generalize the notions of operators in STRIPS planning (see Subsection 2.2.3) and plans in BDI architectures (see Section 2.3.1). We argue that neither single-agent plans nor STRIPS-actions alone are appropriate to model the operational capabilities of an agent acting in multiagent environments. Instead, our concept of operational primitives includes the notions of *patterns of behavior* (PoBs) as well as *joint plans*; it thus allows us to take into account flexible execution procedures on the one hand and coordinated synchronized actions of different agents on the other. Along the lines of Bratman, we stress the strong functional role of operational primitives in mapping beliefs and goals to intentions.

By committing to the execution of OPs that are themselves connected to the agent's current goals, and by the interleaved execution of OPs, the agent's *intentions* are modified. As OPs are hybrid and of different complexity by nature, the intention structure of an agent itself is not homogeneous but is described on different abstraction layers. Following Bratman's *plans as intentions* paradigm, we do not provide an explicit representation of intention, but represent an agent's intention structure at a specific point in time by the current state of execution

- of the agent's active patterns of behavior,
- of its local plans, and
- of the negotiations and the cooperative plans the agent is involved in.

3.2.3 Functional characterization of an agent

The arrows in Figure 3.1 denote functional relationships between the components of the mental model of an agent. They define the flow of control in an agent, mapping its perception into actions. These functional relationships are:

- **Belief generation and belief revision** explain the relationship between the beliefs of an agent and its current perception. They deal with the question of how perception is transformed into beliefs (belief generation), and how existing beliefs change on the basis of perception (belief revision).
- **Situation recognition** extracts (structured) situations from the (unstructured) beliefs of an agent.
- **Goal activation** describes which of the possible goals of an agent are currently options given a set of situations.
- **Planning** maps the agent's current goals into operational primitives for their achievement, i.e., into PoBs, local plans, or negotiation protocols and joint plans. Planning means to decide *what to do.*
- **Scheduling** is the process of merging partial plans for different goals into an execution schedule, which takes compatibilities and (e.g., temporal and precedence) constraints among different subplans into account. Thus, scheduling means to decide *when to do what.*
- **Execution** is responsible for a correct and timely implementation of the commitments determined in the planning and scheduling phase.

The functional processes identified above run in each control layer of the IN-TERRAP architecture. However, their instantiation differs considerably among the layers, which is motivated in the remainder of this section, and which is detailed in the following sections.

Belief generation and belief revision. The generation and revision of the agent's world model, mental model, and social model beliefs are not considered in this book. Models of belief generation, i.e., of deriving knowledge from sensory input, have to take into account the sensory equipment of the agent. A general model would go beyond the scope of this work. We solve the problem of belief generation by making the **symbolic sensor assumption**:

> Sensors provide perception in a symbolic representation that is equal to that of beliefs.

This allows us to omit the problem of how beliefs are generated from perception.

As regards belief revision, we assume that the knowledge representation system provides operators for changing beliefs—adding a proposition, retracting a proposition, or revising a set of beliefs with respect to a proposition—that satisfy the AGM postulates[2] stated by Gärdenfors et al. [Gä88]. However, we do not explicitly describe such a knowledge representation system in this book.

Situation recognition. Situation recognition enables an agent to identify the need for activity. At the behavior-based layer, situations must be identified that require a fast reaction. Thus, situation recognition itself must be fast and efficient. At the local planning layer, situations that require local planning or that in some way affect the agent's current local goals need to be recognized. The cooperative planning layer needs to identify situations whose treatment should involve cooperative planning, such as goal conflicts between different agents, or a potential for cooperation [Had95].

Situation recognition in INTERRAP is an incremental process; time-critical situations are recognized solely based on world model information at the behavior-based layer. If more time is available, it can be improved by investigating possible effects on the agent's local goals or on interactions with the goals of other agents. Thus, incremental situation recognition is an important factor for the integration of reactivity and deliberation.

Goal activation. A situation that the agent has recognized modifies the agent's motivational state: it triggers the emergence of a goal. At the behavior-based layer, this process is hard-wired into PoBs. It can be compared to a rule in a rule-based language: the emergence of a goal corresponds to adding a rule to the conflict set [For82]; the corresponding situation is described by the left-hand side of the rule. In local and cooperative planning, the process of goal activation is more complex; it may involve the explicit construction of a goal state, which is then used by the planning mechanism. In the cooperative case, a goal description is a set of goals for different agents.

---

[2] For a more detailed discussion, we refer to [Neb90] or [Gä88].

Planning. The planning function computes what to do in order to achieve a goal. At the behavior-based layer, this decision is hard-wired in: a goal corresponds to a PoB, and the plan for achieving the goal is given by the procedural body of the PoB. At the local planning layer, planning for a goal involves means–ends analysis and the generation of a local plan whose execution leads to a world state that satisfies the goal description. The generation of a plan can be based either on an axiomatization of the actions in the plan (planning from first principles) or on a plan library (planning from second principles). Finally, cooperative planning means to find a joint plan whose execution satisfies the union of the goals contained in the goal description.

Scheduling. Given a set of goals, and plans how to achieve these goals, scheduling solves the problem of merging these subplans into one executable schedule. Scheduling is problematic because of limitations of the computational and physical resources of the agent, and because of incompatibilities and constraints among plans.

At the behavior-based layer, scheduling means to decide which from the set of *active* PoBs are to be executed in the next control cycle. In particular, the scheduling of PoBs must ensure that those PoBs that react to urgent situations are treated with high priority. At the local planning layer, the scheduling task involves the sequencing of nonlinear plans, the assignment of start and finish times for plan steps in temporally constrained plans, and the recognition of incompatibilities among plans for achieving different goals, causing the planner to modify its plans. A clear distinction between planning and scheduling is not always possible, as it depends on the planning mechanism under consideration. For example, many existing planners incorporate scheduling tasks, too. Finally, at the cooperative planning layer, different concurrent negotiations have to be scheduled.

Execution. At each layer, the agent makes commitments that have to be implemented. The execution function is responsible for initiating the execution of these commitments in time, and for monitoring the execution. For the latter purpose, the execution function interacts with situation recognition: if a commitment has been executed, there are several possible expected results the occurrence of one of which has to be recognized by monitoring changes in the agent's knowledge base.

At the behavior-based layer, the execution function maintains the execution of PoBs, which causes access to the basic actoric, communicative, or sensoric resources. The local planning layer executes primitive plan steps. This is done by activating a procedure, which is modeled as a specific type of a PoB in the behavior-based layer. The cooperative planning layer maintains the execution of (meta-level) negotiation protocols on the one hand, and commitments to (object-level) joint plans, on the other. As the execution of both requires communication with other agents, time-out mechanisms are needed to avoid deadlocks caused by lost or misinterpreted messages.

In summary, the conceptual agent model motivates the design of INTER-RAP, which is presented in the following sections, as a layered BDI approach built on Bratman's original architecture [BIP87]. It defines the mental categories that establish the internal state of an agent (perception, beliefs, situations, goals, operational primitives, and intentions), and the basic functions that modify mental states (situation recognition, goal activation, planning, scheduling, and execution), and that explain the functional relationship between what an agent perceives and what it does.

3.3 The Control Architecture

This section shows how the conceptual agent model is implemented in a practical control framework. It gives an overall description of the INTERRAP architecture, of its basic modules, of the structure of an individual control layer, and of the flow of control between control layers.

3.3.1 Overview

The development of INTERRAP has been guided by the following general design decisions:

- **Layered control:** An agent is described by different levels of abstraction, and of representational and inferential complexities.
- **Layered knowledge base:** The beliefs of an agent are stored in a hierarchical knowledge base; this allows us to restrict the amount and the representation of information available to the lower control layers.
- **Bottom-up activation:** Control is shifted bottom-up; layer i gains control only if layer $i - 1$ is not *competent* to deal with the situation.
- **Top-down execution:** Each layer uses operational primitives defined at the next lower layer to achieve its goals.

The implementation of these design decisions resulted in the INTERRAP agent architecture, which is illustrated in Figure 3.2. An INTERRAP agent consists of three modules: a *World Interface* (WIF), a *Knowledge Base* (KB), and a *Control Unit* (CU). The world interface provides the agent's sensoric, communicative, and actoric links to its environment. The KB and the control unit are structured in three layers. Control layers are the *Behavior-Based Layer* (BBL), the *Local Planning Layer* (LPL), and the *Cooperative Planning Layer* (CPL); each control layer consists of two processes called SG (situation recognition and goal activation) and PS (planning, scheduling, and execution). The knowledge base is partitioned accordingly into a *World Model* (WM), a *Mental Model* (MM), and a *Social Model* (SM).

INTERRAP is a *vertically* layered architecture: control is triggered by the behavior-based layer, which has immediate access to the agent's sensory system, and is shifted upward incrementally until a suitable layer *competent* for

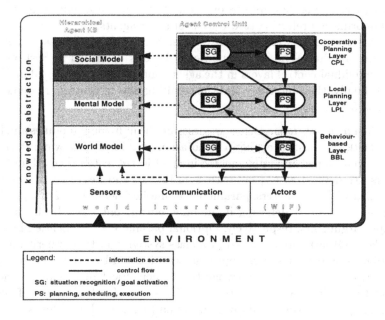

Fig. 3.2. The INTERRAP agent architecture

execution has been found. From there, activity spreads back downward to the behavior-based layer, which is the only layer with direct access to the actoric functions defined in the agent's world interface. This is an important difference to other layered approaches, such as those by Ferguson [Fer92] and Kaelbling [Kae90] where there is concurrent access both to sensory input and actoric output, and where conflicts among the layers have to be resolved by applying appropriate global filtering and suppression mechanisms to ensure that a specific layer only sees those parts of the input that are relevant to it, and to suppress harmful interactions among the decisions of different layers[3].

Each control layer is allowed to access a specific portion of the agent KB; this access is organized incrementally in that each control layer may use information stored in its corresponding KB layer, and in lower KB layers, but is not allowed to access information stored in higher layers. Thus, e.g., the behavior-based layer has access only to the world model part of the KB containing the agent's object-level beliefs about the world. The local planning layer may additionally access the mental model where information relevant for planning is stored such as goal-specific information, plan libraries, or meta-descriptions of patterns of behavior. Finally, the cooperative planning layer may reason about the whole knowledge base including the agent's social

---

[3] In a slightly different context [Min86, p.273ff], Minsky called these mechanisms — that had first been described by Freud — *censors* and *suppressors*.

model, containing descriptions of negotiation protocols, joint plan libraries, and information referring to the goals of other agents.

This section is structured as follows. Section 3.3.2 discusses the world interface module. The agent knowledge base is described in Section 3.3.3. The individual control layers in the agent control unit are outlined in Section 3.3.4; finally, their interplay and the resulting flow of control are depicted in Section 3.3.5.

Notational conventions. Before we describe the individual modules and layers of INTERRAP, we shall explain the notation used in the following: INTERRAP has been designed and implemented in an object-oriented style. Modules and layers are defined as objects with `attributes` and `methods`. Methods can be called with parameters; parameters in method definitions are tagged with one of $+$, $-$, or ?, denoting input only, output only, and both input and output arguments, respectively. Optional parameters are enclosed in square brackets. Objects are arranged in a class system and they inherit methods, attributes, and values from super classes (denoted by the keyword `super`). An object o_1 may send a message to another object o_2 invoking a method m of this object, which is written as $o_2 \leftarrow m$. To denote sets of objects $o_1, \ldots, o_n$, we use the standard set notation $\{o_1, \ldots, o_n\}$ and the list notation $[o_1, \ldots, o_n]$, interchangeably. $[H|T]$ denotes a list with first element H and rest T. Upper case letters denote variables or attribute names; constant and function symbols as well as names of procedures or methods are indicated by lower case letters.

3.3.2 The world interface

The world interface consists of three subsystems implementing the agent's basic facilities for sensing its environment, for performing actions, and for handling the physical aspects of communication with other agents. The individual subsystems of the world interface are briefly discussed in the following:

The sensoric subsystem. A sensor is represented as an object providing methods for sensor calibration, for enabling and disabling sensor activity, and for reading its current value.

```
class sensor
    attributes
            Name
            Value
            Range
    methods
            calibrate +<Name> {...}
            enable +<Name> {...}
            disable +<Name> {...}
            get_val +<Name>{...}
```

The current sensory values are made available to the agent control unit by a *perception buffer* from which the values of sensors can be read, and which can be explicitly updated:

```
class perc_buf
   methods
       init {...}
       clear {...}
       get_val +<Sensorname> {...}
       get_all {...}
       refresh +<Sensorname> {...}
       refresh_all {...}
```

The perception buffer itself calls the methods offered by the individual sensors, e.g., in order to read the current sensory information.

The actoric subsystem. The actoric subsystem controls the physical actions the agent may perform, e.g., motor control routines in the case of a robot. Actors are specified in the following class definition:

```
class actor
   attributes
       Name
       Type                              /* atomic or continuous */
       Range                             /* admissible input values */
   methods
       calibrate +<Name> {...}
       execute +<Name> +<Params> {...}   /* for Type = atomic */
       activate +<Name> +<Params> {...}  /* for Type = continuous */
       suspend +<Id> {...}               /* for Type = continuous */
       deactivate +<Id> {...}            /* for Type = continuous */
```

Here, a distinction is made between two types of actions; atomic actions are actions whose execution is started and then either succeeds or fails. An example for such an action is the classical get_box operator in STRIPS. The execution of continuous actions rather initiates a control process; the process will run until it is explicitly finished, suspended, or deactivated[4]. Examples for these types of actions are actions such as follow_line which activates a control algorithm making a robot follow an induction line on the floor.

The communication subsystem. The communication subsystem provides the functionality of sending messages to and receiving messages from other agents. In accordance with speech act theory [Sea69], sending messages is regarded as a specific type of acting. On the other hand, receiving messages is implemented as a sensing process modeling the agent's receive-queue as a designated sensor. The communication subsystem is described by the following class definitions:

---

[4] This type of action is similar to Firby's notion of a continuous task [Fir94].

```
class send_queue
  super actor /* sending messages is an action */
  methods
      send +<Rcp> [+<Ref>] +<Msg_type> +<Content> {...}

class receive_queue
  super sensor /* receiving messages is a sensing activity */
  methods
      rec_s ?<Sdr> ?<Msg_type> ?<Content> {...}
      rec_a ?<Sdr> ?<Msg_type> ?<Content> [+<Timeout>] {...}
```

It provides two functions allowing an agent to receive messages from other agents: rec_s is a function that waits synchronously until a message has been received. The arguments denote (from left to right) the sender of the message, message type, and actual message content. If the arguments are provided with values, only messages matching the parameter descriptions are returned. rec_a looks for messages asynchronously and fails if no matching messages have been received. rec_a has an optional time-out parameter allowing to specify a time interval during which the message queue is monitored for matching messages. The default value of Timeout is zero.

At the world interface level, messages themselves are represented as tuples

$$\text{Msg} = (\text{Id}, \text{Sdr}, \text{Recp}, \text{Ref}, \text{Type}, \text{Content})$$

where Id is a unique message identifier, Sdr denotes the sender, Recp denotes the recipient, Ref (optional) is a reference to a message-id, Type is one of a list of message types, and Content denotes the actual message content. Admissible forms of Type and Content are given by the definition of the higher-level communication language.

Interfaces. Both perceived information and received messages are sent to the agent knowledge base. The symbolic perception assumption allows us to do this by simply asserting the corresponding facts into the KB. Thus, access to sensory information is provided both upon request from the agent control unit and by automatic KB update by the world interface module. In addition, the world interface may receive commands to execute or activate actors, and to send or receive messages.

3.3.3 The agent knowledge base

The knowledge base of an INTERRAP agent is partitioned into three layers: the world model, the mental model, and the social model. This reflects the structuring of the informational state of the agents, of situations, beliefs, and goals that was defined in Section 3.2. The knowledge of an INTERRAP agent can be represented using the knowledge base AKB (Assertional Knowledge Base), which is described in detail in [Wei95]. In the following, we give a brief summary of the KR mechanism implemented by AKB.

AKB Representation Schema. The basic elements of the knowledge representation schema are the following:

- Concepts (sets of concepts are denoted by $C, C_1, C_2, \ldots$)
- Types (sets of types are denoted by $T, T_1, T_2, \ldots$)
- Attributes (functions $A : C \mapsto T$)
- Features (functions $F : C \mapsto T$)
- Relations $R \subseteq C_1 \times C_2 \ldots \times C_n$

Attributes A may have default values $default(A) = k$; features are attributes of a concept that cannot be changed; $init(F) = k$ denotes the initial value of a feature. Attribute values are typed. Concepts denote classes of individuals; concepts are related to each other by defining relations using the cross-product operator #. An AKB schema declaration is shown in Figure 3.3:

```
[  concept(   name:    ConceptName )
   relation(  name:    RelationName
              domain:  ConceptName₁ # ... # Conceptnameₙ )
   attribute( name:    AttributeName
              concept: ConceptName
              type:    Type )
   default(   name:    AttributeName
              value:   DefaultValue )
   feature(   name:    FeatureName
              concept: ConceptName
              type:    Type
              init:    Init ) ... ]
```

Fig. 3.3. AKB schema declaration

AKB Interface Specification. So far, AKB offers three types of interface services: assertional services, retrieval services, and active information services. Assertional services allow to assert new beliefs into the knowledge base, i.e., to create instances of concepts and relations, and to change the values of attributes of existing concept and relation instances. Figure 3.4 illustrates the available assertional services.

Retrieval services provide access to beliefs that are actually stored in the agent KB. Figure 3.5 shows the retrieval functions currently offered in AKB.

Active information services offer a possibility to access information from the knowledge base upon demand. Requesting an active information service starts a monitoring process that recognizes specific changes in the knowledge base and sends information about these changes to the requesting process, i.e., to specific control layers. The active information services currently available are shown in Figure 3.6.

create_obj(-Id)	returns a unique identification of a newly created KB object.
create_conc(+Id +Concept)	creates an instance of a concept denoted by Concept and binds it to the object identified by Id.
create_rel(+IdList +Rel)	defines an instance of a new relation denoted by relation Rel among the concept instances denoted by the object identifiers in IdList. The ordering of the members of IdList determines their ordering in the relation.
set_val(+Id +Attr +NewVal)	assigns the value denoted by NewVal to the attribute Attr of the concept instance denoted by Id.
del_obj(+Id)	delete an object; deleting an object that is bound to a concept instance deletes the concept instance and all instances of relations having this concept instance as a member.
del_conc(+Id +Concept)	deletes the instance of Concept denoted by Id.
del_rel(+IdList +Rel)	deletes the instance of relation Rel denoted by IdList.
retract_val(+Id +Attr)	removes the value for the attribute Attr of the concept instance denoted by Id.

Fig. 3.4. AKB assertional services

conc_members(+Conc -IdL)	returns a list of all instances of Conc.
conc_memberP(+Id +Conc -Bool)	returns true if the concept instance denoted by Id is a member of Conc.
rel_members(+R -ListOfIdL)	returns a list of list of concept instances denoting all tuples that define relation R.
rel_memberP(+IdL +R -Bool)	Bool returns true if the tuple denoted by IdL is a member of the relation R.
rel_filler(+R +K +IdL$_1$ -IdL$_2$)	for an n-ary relation R, for $1 \leq k \leq n$, and for a list IdL$_1$ = $\{o_1, \ldots, o_{n-1}\}$ of concept instances with $|IdL_1| = n - 1$, rel_filler instantiates IdL$_2$ to $$IdL_2 = \{o | (o_1, \ldots, o_{k-1}, o, o_k, \ldots, o_{n-1}) \in R\}.$$

Fig. 3.5. AKB retrieval services

info_conc(+Conc +Client -Id)	causes any modification of concept instances of Conc to be sent to the address specified in Client.
info_rel(+Rel +Client -Id)	causes any modification of relation instances of Rel to be sent to Client.
info_obj(+OId +Client -Id)	causes any modification of the instance denoted by OId to be sent to Client.
cancel_info(+Id)	Stop information service denoted by Id.

Fig. 3.6. AKB information services

So far, AKB is a prototype knowledge base with a rather rough interface specification: e.g., individual beliefs in the knowledge base are accessed by explicitly referring to the identification of the corresponding KB objects. For the sake of readability, however, in the examples given in this book, we shall assume a higher-level query interface given by a method query of the KB object. A method call kb←query(q) returns the result of query q.

Interfaces. There is a flow of information between the world interface and the KB on the one hand, and between the control unit and the KB on the other. From the world interface, new beliefs derived from the agent's perception are entered into the KB. The problem of belief revision has been discussed in Section 3.2.

The control unit continuously accesses information from the knowledge base. The situation recognition processes at the different layers of INTERRAP evaluate situation descriptions in order to detect new potential goals. The planning process has to evaluate the preconditions of actions when planning. However, the control unit may modify the knowledge base by asserting new beliefs, goals and hypothetical plans. The derivation of new, more abstract beliefs from existing beliefs, which is also called knowledge abstraction, can be anchored either in the knowledge base or in the control part of the agent; whereas the former alternative describes the KB of an agent as an active, blackboard-like system, the latter rather corresponds to the view of a classic AI planning system. Which approach is preferred depends on the power of the inferential services provided by the knowledge base. As the AKB currently only provides very restricted inferential capabilities, in the current INTERRAP system knowledge abstraction has to be done by specific PoBs within the control unit. However, an approach based on a blackboard knowledge base was described in [Mül94], and future work aims at extending the expressiveness of the knowledge base module.

3.3.4 Operation of a control layer

In Section 3.2, it has been argued that the tasks to be performed in each layer of the INTERRAP architecture have similar general characteristics although they have different instantiations in the different control layers. This observation is reflected in the control architecture by the fact that each layer has a uniform structure which is illustrated in Figure 3.7. Each layer implements the basic functions defined in Section 3.2: situation recognition, goal activation, planning, scheduling, and execution. The process SG covers the former two functions, the process PS covers the latter three. Each layer has access to the corresponding part of the knowledge base (see Section 3.3.3); additionally, it may receive messages from the neighboring layers and send messages to them. The interplay among the different layers is described in Section 3.3.5. The operation of an individual layer is implemented in a control cycle as follows:

Generic control cycle. An INTERRAP control layer is defined by a generic object the main method of which describes a sense-recognize-decide-act cycle. In each loop of the cycle, the current beliefs of the agent are scanned for new

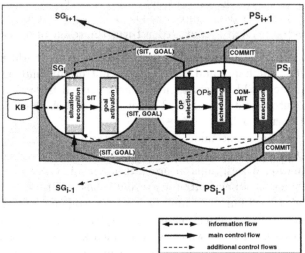

Fig. 3.7. An INTERRAP control layer

relevant situations; these situations and the messages received from the next lower layer are used to compute the agent's new options. These are passed to the planning, scheduling, and execution mechanism, which is responsible for making decisions (i.e., for selecting appropriate operational primitives and for updating the intention structure) and for scheduling and executing the resulting commitments. In particular, this mechanism decides whether it will deal with a specific goal by itself, or whether it will pass the goal up to the next higher control layer. In the former case, the layer decides what commitments to make in order to achieve the goal and schedules the new commitments into the existing intention structure. The execution of commitments that are due to be executed starts at the same time. Execution is linked to the situation recognition process: when an action is executed, a reactor PoB is activated which monitors the expected effects of the action.

Note that the processes of planning and scheduling displayed in Figure 3.7 are interleaved: if the planner has created an intention structure that cannot be scheduled, this is recognized by the scheduler and backtracking may occur enforcing replanning. The interrelationship between planning and situation recognition is discussed below.

The algorithm shown in Figure 3.8 describes the basic activity cycle of an INTERRAP layer. The core of the description of a layer object is the method cycle which defines the basic activity cycle of an INTERRAP control layer, consisting of the above-mentioned steps of belief update (line 20), situation recognition (line 21), goal activation (line 22), generation of situation–goal pairs as input to the planning, scheduling, and execution process (line 23), the competence check splitting the set of situations and goals into a subset for

```
(1)  class layer
(2)  attributes
(3)     Higher, Lower /* neighboring layers */
(4)     Bel, Goals, Ints /* current beliefs, goals, intentions */
(5)     Sit, Sit-desc /* situations and situation descriptions */
(6)     Ops /* operational primitives available */
(7)     Actreq /* activation requests from layer i-1 */
(8)     Com /* commit messages received from layer i+1 */
(9)  [...]
(10) methods
(11)    b-upd +<Bel> {...} /* belief update function */
(12)    s-rec +<Sit> +<Bel> +<Sit-desc> {...} /* sit. recog. */
(13)    g-act +<Sit> +<Goals> {...} /* goal activation fct. */
(14)    comp-check +<Sg> +<Ops> {...} /*competence checking fct.*/
(15)    op-select +<Sg> +<Ops> {...} /* planning function */
(16)    schedule +<Int> +<Ints> {...} /* scheduling function */
(17)    execute +<Int> {...} /* scheduling function */
(18)    [...]
(19) cycle
(20)    Bel = b-upd(Bel);
(21)    Sit = s-rec(Sit ∪ Actreq, Bel, Sit-desc);
(22)    Goals = g-act(Sit, Goals);
(23)    Sg = {(S, G)| S ∈ Sit ∧Gg ∈ Goals ∧ G = g-act(S,_)};
(24)    (Comp-sg,Nocomp-sg)=comp-check(Sg, Ops);
(25)    foreach Sg' ∈ Nocomp-sg /* shift control to higher layer */
(26)      Higher←receive(request, activate(Sg'));
(27)    Int = op-select(Comp-sg, Ops);
(28)    Int = schedule(Int, Com);
(29)    Int = execute(Int);
```

Fig. 3.8. The generic class LAYER

which the layer is competent and for which plans are generated and executed by the layer (lines 27–29), and into another set for which it is not, and for which control is shifted to the next higher layer (lines 25–26).

In Sections 3.4 to 3.6, the general functions for situation recognition, planning, scheduling, or execution shall be replaced by specific ones for the different control layers. From the point of view of software engineering, this can be done very elegantly by defining the three control layers as special layer objects that inherit the attributes and the main loop from the general INTERRAP layer, and which complement the general description of a layer by their specialized methods.

The separability problem revisited. The structuring of a control layer into two processes as shown in Figure 3.7 leads to an issue discussed in control theory (see Section 2.2.1), namely that of the separability of situation recognition and goal activation (SG, called *state estimation* in control theory) on the one hand, and planning, scheduling, and execution (PS, called *input regulation* in the control-theoretic case), on the other. The separation property holds for the two processes if the decisions made by the PS have no adverse effects on the ability of the SG process to recognize the current state of the system, and if the situation recognition has no adverse effect on the ability of the PS to control the behavior of the agent.

In [DW91, p.22ff], Dean and Wellman argued that in designing planning systems, this property cannot automatically be assumed, as the quality of plans often depends on the quality of situation recognition and the cost of situation recognition has to be weighed against its profit. Thus, robots may have special sensors whose operation is expensive (e.g., in terms of energy consumption). If information is lacking, the planner itself may decide to activate the sensor if the information is important enough. On the other hand, if a robot is designed to orient itself by using induction lines on the floor, and the planner decides to leave these lines for some reasons, this clearly interferes with the ability of situation recognition to operate.

INTERRAP offers two additional links between situation recognition and planning to deal with these problems: firstly, there is a link from the execution module to situation recognition; it serves to enable or disable the monitoring of specific situations that are derived e.g., from the monitoring conditions of PoBs, or from the expected effects of procedures called by the local planning layer. Note that the process PS at layer i can enable and disable monitors both in SG_i and in the next lower layer SG_{i-1}. Enabling and disabling monitoring functions is an important tool when it comes to avoiding foreseeable adverse interrelationships between PS and SG (see Section 4.2.4). Secondly, there is a link from planning to situation recognition. This link allows the planner to order additional information upon demand from the SG process, whenever this is necessary to make planning decisions.

3.3.5 The flow of control

The behavior of an INTERRAP agent results from the interplay among the individual control layers. Two basic control directions are bottom-up control and top-down control. They determine three generic control paths describing three types of control flows in an agent for dealing with different classes of problems or tasks. This section includes a discussion of both top-down and bottom-up control flow; the generic paths of control through an INTERRAP agent are illustrated. Finally, the section surveys the possible interactions among different layers.

Bottom-up control: upward activation requests. The initiation of activity in INTERRAP is handled in a bottom-up fashion using the upward activation request mechanism. New situations are recognized by the process SG_i at layer i, and the planning, scheduling, and execution process PS_i at this layer is activated. If PS_i decides not to be competent for dealing with the corresponding situation-goal pair, it sends an activation request to the situation recognition process SG_{i+1} at layer $i + 1$; there, the description provided by layer i is enhanced by additional information that is available to layer $i + 1$ and that is required to produce a suitable goal description for PS_{i+1}. The result of the activation request is finally reported back to PS_i.

The mechanism implements a competence-based control flow; it ensures that situations requiring quick reaction are handled by the behavior-based layer, whereas other situations that encode more complex planning problems are shifted upward until a layer has been reached which is competent for solving the problem. The advantage of this bottom-up approach is that lower layers do not need to know about the capabilities of higher layers: All they have to decide is whether they are able to solve a specific problem for themselves. For instance, at the behavior-based layer, this can be done quickly and efficiently based on simple matching and table lookup mechanisms; this will be spelled out in more detail in Section 3.4.

A problem occurs in case a situation cannot be dealt with by any layer; this, however, implies that the agent is faced with a task it has not been designed to solve. Thus, if the cooperative planning layer does not declare itself competent for dealing with a situation, it has to report this down to the requesting layer; this process is iterated until the layer that originally initiated activity receives the failure report. In this case, either the agent has to react by sending a decline message (in case activity was caused by a task request by another agent), or by some emergency routine (in case the situation is e.g., a conflict situation the agent is not familiar with) which will allow it to deal with the situation *somehow* (even if probably rather ineffectively). For example, an emergency routine allowing forklift agents to resolve blocking conflicts in many cases where no other mechanism works is to perform a couple of random moves; alternatively, the agent might request help from other agents in these situations.

Top-down control: downward commitment posting. The second type of control flow through the agent determines the actions it performs in order to deal with certain situations. Whereas the activation of layers is done bottom-up, acting is organized in a top-down manner. The planning and scheduling processes of neighboring layers coordinate their activities by communicating commitments. This communication is directed from higher layers to lower layers. Commitments posted from the cooperative planning layer to the local planning layer are partial plans that have been devised by the cooperative planning layer during a joint plan negotiation and that describe the role of the agent in the joint plan. These commitments are incorporated into the local schedule maintained at the local planning layer. On the other hand, the local planning layer commits itself to the execution of procedure PoBs; these commitments are posted to the behavior-based layer and cause the activation of the corresponding procedures. Commitments at the behavior-based layer itself cause the execution of actions defined in the agent's world interface. Similar to top-down control, after posting a commitment down to layer $i-1$, layer i will wait for an acknowledgment saying whether the commitment could be executed successfully or whether it failed. Based on the reports of the lower layer, the higher layer continues in executing its plan or has to replan, respectively.

Generic paths of control. Due to the bottom-up control mechanism activity is always initiated by the behavior-based layer. Starting from there, there are three generic control paths that are illustrated in Figure (3.9.a–c). Figure (3.9.a) describes the *reactive path*: emergency situations or situations that can be dealt with by routine PoBs without requiring explicit planning are recognized and handled at the behavior-based layer. If the agent has to tackle more complex situations for which there is no suitable PoB, control takes the *local planning path*. A typical class of situations that are handled by using this path are situations requiring local task planning, e.g., planning a transportation order. For this class of situations, control is shifted upward to the local planning layer; there, a plan is generated and executed by activating sequences of procedure PoBs (see Figure (3.9.b)). The *cooperative path* additionally involves the cooperative planning layer; problems whose solution exceeds the problem-solving capabilities of the local planning layer as they require coordination with other agents are shifted upward and handled by devising a joint plan. In executing the joint plan, the corresponding partial plan for the agent is posted downward to the local planning layer, from where procedure PoBs are activated. The processes in Figures (3.9.a–c) are generic

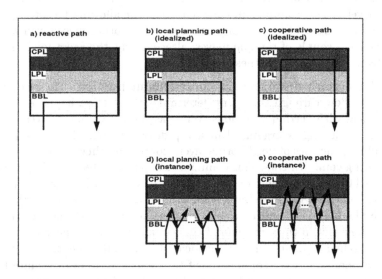

Fig. 3.9. Generic control paths

in that they describe idealized versions of control flow.

Often, the activities occurring at the different layers are not clearly separated nor do they follow the strict temporal ordering implied by Figures (3.9.a–c). Rather, control switches between different layers during processing. Figure (3.9.d) shows the interleaving of planning and execution; planning is an ongoing incremental process, and future planning decisions depend on the outcome of current PoB calls. Analogously, Figure (3.9.e) implies that there is a close connection between cooperative and local problem solving. For example, in order to evaluate a plan that has been proposed by another agent, the cooperative layer needs to evaluate its role in the plan, which is a functionality offered by the local planning layer. cooperation may also be an option for local planning. These issues will be detailed by the application example of the loading dock in Chapter 4.

Additional inter-layer coordination. INTERRAP offers additional types of messages exchanged between control layers. There are two cases to be considered: communication between the behavior-based layer and the local planning layer, and communication between the local and the cooperative planning layer.

Communication between BBL and LPL. Important interactions between the behavior-based layer and the local planning layer include enabling and disabling of patterns of behavior from the local planning layer as well as the context-dependent inhibition and permission of patterns of behaviors in specific situations. These messages can be used by the local planner as mechanisms to control the activity of the behavior-based layer; they are thus a means to increase coherence. A further important interface functionality between the local planning layer and the behavior-based layer is that the former layer must be able to stop the execution of a pattern of behavior, e.g., in case the current plan has to be changed. Possible messages sent from the behavior-based layer to the local planning layer include the request to devise a plan for a goal situation pair, to evaluate or to interpret a given plan, and to stop activity regarding an earlier request. As these messages are layer-specific, they are discussed in more detail in the following sections where the individual layers are described.

Communication between LPL and CPL. The local planning layer may receive from the cooperative planning layer requests to interpret a plan (which is the single-agent projection of a joint plan), to evaluate a single-agent plan, or to stop activity as regards an earlier request. Furthermore, the local planning layer may request from the cooperative planning layer to devise a plan for a given situation-goal description, and to evaluate and to interpret a joint plan, which may have been proposed by another agent. As above, the individual messages shall be explained in the respective sections dealing with the different control layers.

3.4 The Behavior-Based Layer

The behavior-based layer serves two purposes: firstly, it incorporates the reactive abilities of the agent, which allow it to deal in real-time with a class of emergency situations. Secondly, it provides the agent's procedural knowledge needed to perform routine tasks efficiently. The behavior-based layer interacts with the local planning layer by executing procedure calls from the local planning layer, and by activating the local planning layer in case situations occur that cannot be dealt with by the behavior-based layer itself.

3.4.1 Overview

The behavior-based layer has to recognize emergency situations and to respond quickly to these situations; it also needs to schedule the execution of fairly complex procedures. The main ideas and concepts underlying the behavior-based layer are the following:

- The part of the knowledge base accessible to the behavior-based layer is the world model; this means that situation recognition can be performed on a set of ground facts, for which efficient matching algorithms are available.
- The number of situations to be recognized is limited by the set of situation descriptions that have to be monitored by the behavior-based layer. This together with the propositional knowledge representation allows an efficient implementation of situation recognition.
- There is a hard-wired link between reactor goals and situations recognized on the world model; recognizing a situation directly leads to the emergence of a reactor goal; thus, goal activation can be done efficiently.
- The mapping from situation–goal pairs to intentions is achieved by means of *reactor PoBs* that provide a hard-wired connection between situations and executable programs. Thus, planning is restricted to situation-driven decision-making with a short lookahead.
- Specific situation–goal pairs that are recognized in the situation recognition part of the behavior-based layer, but that are not matched by a reactor pattern cause an upward activation request to the local planning layer.
- Commitments to the execution of procedures from the local planning layer cause the activation of procedure PoBs at the behavior-based layer.

Thus, reactivity in the behavior-based layer is achieved by restricting the amount and the representation of the beliefs, by restricting the power of situation recognition, and by providing a hard-wired connection between situation and action. However, to ensure reactivity, additional problems have to be solved. Firstly, the number of different situations that have to be recognized in the behavior-based layer must be restricted effectively. Secondly, due to the resource-boundedness of the agent, it may not be able to pursue all patterns of behavior (i.e., all reactor goals the agent has at a certain time) in parallel, but rather needs to select those with the highest priority. This meta-reasoning, however, can be a source of time-consuming complexity by itself; thus, mechanisms are required to restrict its complexity (*resource-bounded computation*). Thirdly, reactivity requires the ability to interrupt the execution of PoBs at specific points in time. Fast situation recognition does not help if the procedures that are actually used are monolithic and cannot be interrupted in case something more important happens.

In the remainder of this section, the solution to these problems is described. In Section 3.4.2, the different types of patterns of behavior are presented; their operational semantics is defined. Section 3.4.3 then describes how the processes of situation recognition and goal activation on the one hand, and of planning, scheduling, and execution on the other are modeled in the behavior-based layer by an instantiation of the general layer cycle defined in Figure 3.8. Section 3.4.4 discusses the interface to the local planning layer.

3.4.2 Patterns of behavior

Patterns of behavior are the essential data structures of the behavior-based layer. There are two types of PoBs, i.e., reactor PoBs and procedure PoBs, which differ mainly with respect to their activation: reactors are triggered by external events that cause changes in the agent knowledge base; they are used to make the agent reactive to unforeseen events. Procedures are activated by posting down commitment requests from the local planning layer; at the behavior-based layer, they establish a commitment to executing a routine procedure.

Structure of PoBs. Figure 3.10 shows the definition of a PoB as a data structure in INTERRAP. It consists of two parts: a description part and an execution part. The description part includes the attributes and methods relating to the activation and monitoring conditions; it is used by the control cycle of the behavior-based layer (see Section 3.4.3) to maintain and to control the PoBs. Additionally, it can be used by the local planning layer to make predictions about potential interactions between reactor PoBs and the current plans of the agent (see Section 3.5). The execution part contains the actual body of the PoB, which is a program that is executed if the pattern of behavior becomes selected for execution. In the following, the individual attributes and methods characterizing a PoB are explained in more detail. The operational semantics of PoBs is defined by the layer control cycle. It is discussed in Section 3.4.3.

Activation and commitment. A PoB can have three different states: it can be inactive, it can be the goal of the agent (then, it is active), or the agent can be committed to it. The status transition of a goal from inactive to active is called the **activation** of the PoB. The two PoB types *reactor* and *procedure* mainly differ in their activation. Whereas procedures are explicitly

```
class PoB
  attributes
    Name
    Type              /* ∈ { reactor, procedure } */
    Args              /* arguments of PoB activation
    State             /* ∈ { inactive, active, committed } */
    Resources         /* list with needed resources */
    Priority          /* integer: static priority of the PoB */
    Act-cond          /* activation condition */
    Succ-cond         /* successful termination condition */
    Fail-cond         /* failure condition */
    Exception-list    /* list of exception descriptions */
    Body              /* executable program */

  methods
  [...]
    meth init             /* initialize PoB */
    meth activate         /* activate PoB */
    meth step             /* execute next step of PoB */
    meth uncommit         /* stop execution of PoB */
    meth check-act-cond   /* check activation condition */
    meth check-monitors   /* check monitoring conditions */
    meth enable           /* enable activation of PoB class */
    meth disable          /* disable activation of PoB class */
    meth inhibit <Sit>    /* inhibit activation */
    meth permit <Sit>     /* cancel preceding inhibition */
```

Fig. 3.10. Patterns of behavior (PoBs)

activated by call from the local planning layer or from another PoB, reactors are activated if their activation condition (which is specified under the attribute Act-cond) is valid with respect to the current state of the agent's world model beliefs. In the following, let $\mathcal{B} = \mathcal{B}_r \cup \mathcal{B}_p$ denote the set of PoBs, where $\mathcal{B}_r$ are the reactor PoBs and $\mathcal{B}_p$ are the procedure PoBs in $\mathcal{B}$. We now define the notion of *active PoBs*.

Definition 3.4.1. *A reactor PoB* r $\in \mathcal{B}_r$ *is active at time* t_i *if either:*

– *the activation condition* r←Act-cond *is satisfied at time* t_i, *or*
– r←Act-cond *was satisfied at time* t_j, $j < i$, *and there is no* $j < k \leq i$ *such that either the termination condition* r←Succ-cond *or the failure condition* r←Fail-cond *of* r *were satisfied at* t_k.

A procedure PoB p $\in \mathcal{B}_p$ *is active at time* t_i *if either:*

– *a commitment message* commit(p, args) *was received from the local planning layer at time* t_{i-1} , *or*
– p *was activated by the execution of the instruction* activate(p, args) *from within the body of a committed PoB* q *at time* t_{i-1}.

- p *was activated at time* t_j, $j < i$, *and there is no* $j < k \leq i$ *such that either the termination condition* p←Succ-cond *or the failure condition* p←Fail-cond *of* p *were satisfied at* t_k, *or a message* uncommit(p) *was received from the local planning layer at* t_k.

Furthermore, reactors differ from procedures with regard to their termination. Whereas a reactor is only ended either if its termination condition or its failure condition is satisfied, a procedure b can be terminated if the next higher layer (the local planning layer) is no longer committed to b.

Thus, several PoBs may be active at a certain time; however, it may not be possible for the agent to pursue all active PoBs at a time, as they may access shared resources (sensors or actuators) during their execution, or as they may be otherwise semantically conflicting. Whereas the latter type of restriction has to be validated at design time by the system designer (it does not make much sense to concurrently execute two PoBs, one for moving to landmark l_1 and another for moving to landmark l_2), the former inconsistency can be checked at runtime: each pattern of behavior has the attribute Resources <resource-list> whose value <resource-list> is the list of resources that are used by the PoB. This information allows a simple compatibility check: given two PoBs b_1, b_2, we can define a symmetrical predicate $compatible(b_1, b_2)$ if $b_1.Resources \cap b_2.Resources = \emptyset$. Using this predicate, we can define the notion of PoBs to which the agent is actually committed at time t_i. For this purpose, we first introduce the concept of a maximal compatible set.

Definition 3.4.2 (maximal compatible subset). *A set* $\emptyset \neq B \subseteq \mathcal{B}$ *of active PoBs is a maximal compatible subset of* $\mathcal{B} \neq \emptyset$ *at time* t_i, *written* $mcs(B, \mathcal{B}, t_i)$ *if, for each* $b \in B$ *we have:*

- *for each* $b' \in B$, *we have* $compatible(b, b')$ *and*
- *for each* $b'' \in \mathcal{B} - B$ *we have:*
    - $\neg compatible(b, b'') \Rightarrow$ *(b.Priority* $\geq b''$.Priority *or ex.* $\hat{b} \in B$ *such that* $\hat{b}.Priority > b''.Priority$*)*
    - $compatible(b, b'') \Rightarrow$ *ex.* $\bar{b} \in B$ *such that* $\bar{b}.Priority > b''.Priority$

Here, b.Priority denotes the static priority of b. *For any* $b \in \mathcal{B}_r$, $b' \in \mathcal{B}_p$, *we require that b.Priority* $> b'.Priority$.

For example, if $\mathcal{B} = \{b_1, b_2, b_3\}$ with $b_1.Priority = 1$, $b_2.Priority = 2$, $b_3.Priority = 3$, given that $\neg compatible(b_1, b_2)$ and $\neg compatible(b_2, b_3)$, then $B = \{b_3, b_1\}$ is a maximal compatible subset of $\mathcal{B}$. Note that Definition 3.4.2 is not unique in case the ordering induced by the priority is only partial. E.g., for $\mathcal{B} = \{b_1, b_2, b_3, b_4\}$ with $b_1.Priority = 3$, $b_2.Priority = b_3.Priority = 2$, $b_3.Priority = 1$, given the constraints $\neg compatible(b_2, b_3)$ and $\neg compatible(b_2, b_4)$, both $B = \{b_1, b_2\}$ and $B' = \{b_1, b_3, b_4\}$ are maximal compatible subsets of $\mathcal{B}$.

```
1    /* Input B is a set of PoBs written in list notation */
2    func MCS(B)
3    {
4      B' = sort(B, priority, >); /* sort B descendingly by priority */
5      return MCS(B', []); /* second argument is accumulator */
6    }
7  /* Input Acc is a list of PoBs (used as an accumulator) */
8    func MCS(B, Acc)
9  {
10     if B = [] then return Acc; /* end of recursion */
11     else
12       B = [H|T];
13       if compatible(H, Acc) then
14         return MCS(T, append(Acc, H)); /* incorporate H */
15       else
16         return MCS(T, Acc); /* do not incorporate H */
17       fi
18     fi
19 }
```

Fig. 3.11. The algorithm MCS

Figure 3.11 shows the definition of the recursive algorithm MCS, which computes a maximal compatible subset of a set of PoBs. Next, we show that algorithm MCS terminates and computes a maximal compatible set of PoBs.

Theorem 3.4.1. *Let t be a point in time; let B be a finite input set of active PoBs at time t; then, the following properties of algorithm MCS hold:*

1. *Algorithm $MCS(B)$ terminates.*
2. *The result B' of algorithm $MCS(B)$ is a maximal compatible subset of B, i.e., $mcs(B', B, t)$. As t is fix, we write $mcs(B', B)$.*

The proofs of all theorems in this book are collected in Appendix A.

The notion of resource compatibility defined above is coarse in that it does not allow concurrent commitments to two PoBs that *could* access the same non-sharable resources, even if the actual periods of resource consumption do not overlap. A more sophisticated strategy which might allow PoBs to lock individual resources dynamically, i.e, only for the period they are actually accessed, is subject to future work.

The set of intentions maintained at the behavior-based layer at time t_i is defined by the set of PoBs whose status is committed. These are executed at time t_i. All other PoBs that are active but not committed remain active and are checked again at time t_{i+1}, in the next control cycle.

Execution of PoBs. Executing a PoB means executing the program that constitutes its body. As this body of a PoB can be a complex routine, its execution has a duration that must not be neglected. In particular, while executing a PoB, events may occur to which the agent must react. Therefore, an important requirement for body programs is that they are interruptible.

We satisfy this requirement by looking upon a program as consisting of a sequence of primitive atomic instructions, and by defining a step-wise execution procedure. After each execution step, it can be decided again whether to interrupt the program or whether to continue its execution.

Another critical topic is the granularity of execution: it must not be too coarse in order to keep reactivity of the behavior-based layer, nor must it be too fine in order to reduce the overhead for changing between the execution frames of different PoBs too often. So far, this is not supported by the system: the agent designer herself is responsible for selecting an appropriate granularity when programming the bodies of PoBs. In summary, an execution language for PoBs needs to satisfy the following requirements:

- It has to allow stepwise execution providing reasonable stepwidths and allowing to specify *atomic* activities that must not be interrupted.
- Its primitives should be the activation of WIF primitives, such as actions and sensors, and calls to the local planning layer.
- It shall offer language constructs such as composition, tests, and iteration. For determining the values of test and iteration predicates, access to the knowledge base is needed.
- Since we require that PoBs may be compiled down from plans[5], they will activate other PoBs. Therefore, sequential and parallel activation should be supported.

Figure 3.12 shows the EBNF syntax of a language fulfilling these requirements. Keywords (e.g., **while, if**) appear in bold face, primitives in italics.

```
program        ::=  block [';' program]
block          ::=  '{' block-content '}' | primitive-inst
block-content::=    while condition do block-content od |
                    if condition then block-content else block-content fi
                    | primitive-instruction
primitive-inst=     wif-execution | lpl-call | pob-activation | modifier
condition      ::=  atomic formula
wif-execution::=    ex '(' wif-primitive ')'
lpl-call       ::=  call '(' lpl ',' do '(' goal-spec ')' ')'|...
pob-activation=     activate '(' pob-spec-list ')'
wif-primitive::=    list of domain-dependent actuator/sensor calls
goal-spec      ::=  formula
pob-spec-list::=    '[' { pob-spec }+ ']'
pob-spec       ::=  pobname '(' parameter-list ')'
pobname        ::=  atom
modifier       ::=  assert ground formula | retract ground formula
```

Fig. 3.12. Execution language for PoBs (EBNF syntax)

---

[5] The implementation of this requirement, however, goes beyond the scope of this book.

Dots denote incomplete definitions. In the following, we shall define the semantics of the execution language.

Semantics. The operational semantics of the language presented above is defined by the method *step* which takes as argument a language expression. Figure 3.13 shows the recursive definition of *step*. It is described using a meta language. *Ex*, *call*, and *activate* denote the physical actions of executing a world interface primitive, calling the local planning layer, or activating a pattern of behavior. *While do od* and *if then else fi* are interpreted as usual. In order to distinguish the meta language constructs from their object-level counterparts, we do not use bold type for the former ones.

```
step({ P })                         ≝   block = true; step(P);
                                        block = false

step(P; Q)                          ≝   step(P); step(Q)

step(while c do P od)               ≝   if c then step(P; while c do P)
                                        else true fi

step(if c then P else Q fi)         ≝   if c then step(P) else step(Q)

step(ex(P))                         ≝   ex(P); if block = false then
                                            exit fi
                                        /* exit:leave execution */

step(call(lpl, X))                  ≝   lpl←request(activate(X)) ;
                                        if block = false then exit
                                            else true fi

step(activate([P|Rest]))            ≝   activate(P); step(activate(Rest);
                                        if block = false then exit
                                            else true fi

step(activate([]))                  ≝   true
```

Fig. 3.13. The *step* function

The function *step* is used to implement the method **step** of a PoB which is called by the control cycle. Upon activation, the method **step** calls the function *step* with the body of the PoB as its argument, or, if this call has been initiated earlier, performs the next recursive call of *step*.

The role of the *block* construct which is denoted by brackets is to provide a simple transaction mechanism: the program part in brackets is an atomic transaction which is executed without allowing another PoB to interrupt it. Thus, the opening bracket can be looked upon as starting a semaphore, while the closing bracket ends the semaphore.

Additional work has to be done before and after each execution step: the local environment of a PoB, such as its local variable values, has to be restored each time the state of the PoB changes from **active** to **committed**; the environment has to be saved each time the state of the PoB changes from **committed** to **active**.

Monitoring the execution. As soon as a PoB becomes active, a set of conditions have to be monitored. This is necessary e.g., to determine when the PoB should be terminated as it has achieved its purpose or as it has failed. Monitoring is initiated by calling the check-monitors function, which evaluates (i) the termination condition specified in the Succ-cond attribute, (ii) the failure condition specified in the Fail-cond attribute, and (iii) all user-defined exceptions specified in the attribute Exception-list (see Figure 3.10). An exception is a tuple of the form (Cond, CleanUp), where Cond specifies the actual exception condition to be monitored and CleanUp specifies certain activities that have to be performed if the exception occurs. Termination and failure conditions can be regarded as specific exceptions with a fixed semantics. The use of specific clean-up tasks is known from AI planning [Fir94], where it is used to ensure a regular termination of a task and to produce a consistent state after the task has ended, especially in case of a failure.

The internal treatment of monitoring PoBs is based on the observation that exceptions can be looked upon as special PoBs where the exception condition corresponds to the activation condition and the clean-up condition corresponds to the body program. Therefore, the internal processing of the monitoring is as follows: when the description of a user-defined PoB is loaded (which happens at the time the agent is created), a reactor PoB is created internally for each monitoring condition of a user-defined PoB. If the PoB is activated, the reactors corresponding to its monitoring conditions are *enabled*; as soon as the PoB terminates, they are *disabled* (see below).

Special functions. Apart from the general functions controlling the activation, the execution, and the monitoring of PoBs, a variety of additional methods are defined for a PoB object. They are briefly discussed in the following.

Aborting execution. The method uncommit can be used by the local planning layer to abort an active procedure. The semantics of uncommit is that the failure condition of the PoB is set to true by adding 'V true' to the original condition. Thus, the PoB fails and the clean-up task corresponding to the failure condition is executed in order to produce a consistent result state.

Enabling and disabling PoBs. The methods enable and disable can be used to modify the set of PoBs that are monitored at run-time. If the situation rules out the possibility that a specific PoB *might* become active, the activation condition of this PoB will no longer have to be checked. E.g., a robot that is not driving does not need to check for a collision. PoBs can also be disabled if their activation would lead to conflicts with other activities, and can be enabled again if this potential conflict is no longer present. Enable and disable can be called both from the local planning layer and from within the bodies of other PoBs. They provide a means to ensure coherent behavior of the agent. However, using disable and enable for these purposes often does not provide the appropriate granularity, as it is often intended to enforce a

PoB to reveal a certain behavior (e.g., not to move to specific places while trying to dodge another agent) rather than to disable the PoB as a whole (i.e., not to dodge at all). For an example, see Section 4.2.4).

Context-dependent inhibition. Two methods inhibit and permit are available to restrict the output of PoBs in a context-dependent manner. The method inhibit is called with an argument Sit, which is a formula. The operational semantics of the method call inhibit(Pob,Sit) is that the termination condition of Pob is enhanced by 'V Sit'. Thus, whenever a situation Sit occurs, the termination condition of Pob evaluates to true and Pob is terminated. Inhibitions can be cancelled by calling the method permit(Pob,Sit); in this case, the appendix 'V Sit' is removed from the termination condition. This removal is merely based on a syntactical check whether Succ-cond contains the string denoted by 'V Sit' rather than checking whether Sit is a logical consequence of Succ-cond.

The inhibition mechanism provided by INTERRAP is very similar to that proposed by Brooks in his subsumption architecture (see Section 2.4). The difference is that our inhibition mechanism is much more expressive since inhibition is hard-wired into Brooks' behavioral hierarchy whereas inhibition of PoBs is subject to reasoning that can be performed in the local planning layer. In so far, the mechanism described here is more similar to that proposed by Dabija [Dab93].

In summary, the concept of PoBs provides a flexible basis for programming reactive agent behavior and procedural skills. In the following section, we describe how PoBs are maintained by a reactive control mechanism.

3.4.3 The control cycle

The behavior-based layer is an instance of the generic class *layer* (see Section 3.3.4). It defines the operational semantics of PoBs. It instantiates the five basic functions illustrated in Figure 3.7: situation recognition, goal activation, planning, scheduling, and execution.

Phases of the control cycle. Figure 3.14 shows the class definition of the behavior-based layer. The control cycle is a method of the class BBL (line 16 ff). In each loop of the cycle, first the perception is updated by calling the corresponding perceptual methods in the world interface (see Section 3.3.2); then, new messages from the local planning layer are looked up. The world model beliefs of the agent are updated using both the perception and the messages. The two foreach loops (lines 20–37) implement situation recognition and goal activation in the behavior-based layer. Firstly, those PoBs are checked that are currently active. According to Definition 3.4.1, it is first tested whether the PoB has produced any exceptions. They are treated by ending the corresponding behavior (in case of termination or failure) and by activating the corresponding clean-up PoBs (line 28). Secondly, PoBs are checked that have been inactive before. In the case of reactor PoBs, the activation conditions are tested; for procedures, it is looked up whether a `commit` message has been received from the local planning layer. In case this is true, the PoB is activated, and its monitoring conditions are enabled (line 35).

At this stage of the cycle, the variable `B-goals` contains the set of currently active PoBs. Planning at the behavior-based layer is achieved by the hard-wired link between the description part of a PoB to its executable body. Thus, looking at a PoB as a situation–action rule $s \to a$, given an instantiation $\theta(s)$ of the situation (which is the activation condition or the commitment message received by the local planning layer, respectively) with a substitution θ, the decision as to what to do is given by the instantiation of the body $\theta(a)$ of the PoB. The next step is scheduling. At the behavior-based layer, scheduling active PoBs means determining to which of the PoBs the agent should commit itself, and which ones it should defer to a later point in time. Scheduling is achieved by the method `mcs` (line 38). This method implements algorithm MCS based on Definition 3.4.2. It returns a maximal compatible subset of the set of active PoBs. Note that an important implication of Definition 3.4.2 is that reactor PoBs are scheduled before procedure PoBs. Thus, whenever an agent has to react to a new situation while executing a procedure b, the corresponding reactor r is treated with a higher priority: if b and r are not compatible (see 3.4.2), it is guaranteed that r is selected by the scheduler for execution.

The scheduling function ends the process of intention formation. At this stage of the control cycle, a maximal set of compatible PoBs has been selected that can be executed concurrently. Execution is done by activating the method `step` of the selected PoBs which executes the next part of the body of the PoB (see Section 3.4.2 for the definition of this function).

To summarize, the functions of an INTERRAP layer are modeled as follows:

- **Situation recognition:** PoBs have *activation conditions* (for reactors), which trigger their activation, and *monitoring conditions*, which monitor success, failure, or exceptions during execution. These conditions correspond to relevant situations that must be recognized by the behavior-based layer.
- **Goal activation:** The activation of a PoB implies the activation of a reactor goal (in case of a reactor) or a procedure goal (in case of a procedure).
- **Planning:** The plan used to satisfy a PoB is provided by the execution body of the PoB. Thus, planning in the sense of deciding *what to do* is a one-step decision process at the behavior-based layer.
- **Scheduling:** Scheduling is to coordinate the execution of different active PoBs. Two mechanisms are used for this purpose: static priorities and dynamic detection of resource access conflicts.
- **Execution:** At each point in time, there is a set of concurrently executable PoBs as a result of the scheduling process. These are the agent's intentions

```
 1 class BBL
 2 super layer
 3   attributes
 4     Poblist, B-goals, Inact-pob /* all/active/inactive PoBs */
 5     Perc /* current perception */
 6     Int /* list of currently pursued PoBs */
 7     [...]
 8   methods
 9   [...]
10     meth perc-upd
11     meth upd-lpl-msg
12     meth bb-upd(+<Bel>, +<Perc>)
13     meth enable-monitors(+<PoB>)
14     meth disable-monitors(+<PoB>)
15     meth mcs(+<Pobset>)
16     meth cycle
17       Perc = perc-upd;
18       Msgs = upd-lpl-msg;
19       B-bel = bb-upd(B-bel, Perc, Msgs)
20       foreach B ∈ B-goals do
21         Exceptions = B←check-monitors;
22         foreach (M,C) ∈ Exceptions do
23           if M ∈ { Succ-cond, Fail-cond, Clean-up} then
24             B-goals = B-goals - {M};
25             Inact-pob = Inact-pob + {M};
26             disable-monitors(B);
27           fi
28           C←activate;
29         od
30       od
31       foreach B ∈ Inact-pob do
32         if B←check-act-cond ∨ exists msg(commit(B, Args)) then
33           B-goals = B-goals + {B};
34           Inact-pob = Inact-pob - {B};
35           enable-monitors(B);
36         fi
37       od
38       Int = B-goals;
39       foreach I ∈ Int do
40         I←step od
```

Fig. 3.14. The control cycle of the behavior-based layer

that are executed. An important requirement is that execution has to be interruptible, i.e., that there is a stepwise execution mechanism.

Reactivity. A main requirement stated at the beginning of this section has been that the behavior-based layer should allow the agent to react quickly to unforeseen situations. In this paragraph, we briefly discuss how the design of the layer and the control cycle affects the implementation of the reactivity requirement.

Situation recognition and reactivity. A crucial precondition of being able to react to a situation is to recognize it in time. The complexity of the situation recognition problem is given by the number of reactor PoBs whose activation conditions must be checked and by the number of commitment messages received by the local planning layer. Whereas the latter is not critical if we assume that planning time is long compared to reaction time, the former may become a problem when the number of reactor PoBs to be monitored is high.

There are three ways of restricting the complexity of situation recognition, two of which have been discussed before in the context of knowledge representation: one way is the restriction of the world model to ground, atomic facts. This corresponds to a restriction of both the representation and the amount of the knowledge the layer has to reason about. The second way is the restriction of the expressiveness of the mechanism for checking the truth of a precondition to simple pattern matching against the knowledge base. These two restrictions determine how fast an individual situation can be recognized. The third way of keeping situation recognition manageable is to restrict the number of PoBs that have to be monitored by a *focusing mechanism.* The focusing mechanism used in INTERRAP is based on that proposed by Dabija [Dab93]. The main idea is that reactors are context-sensitive: some situations are expected to occur only in specific situations. In a different context, where these situations are not expected to occur, the corresponding PoBs can be disabled by the local planning layer. For example, a robot only needs to watch out for a potential collision with an obstacle while it is driving. As soon as it stops driving, the obstacle avoidance PoB may be disabled.

Commitment and reactivity. Having determined an agent's options, the next important factor influencing its ability to react is how fast it can *decide* which of its options to pursue. The hard-wired link between goals and execution procedures provided by PoBs is one aspect of making this decision efficient. The second aspect is given by how fast the agent decides which PoBs to pursue and which ones to defer for reasons of resource-boundedness.

Currently, there are two mechanisms supporting this process of decision-making. One is to allow the designer of an agent to pre-order PoBs by assigning a static priority to them. The selection algorithm (see Definition 3.4.2) guarantees that PoBs with higher priorities are preferred to those with lower priorities. Especially, reactors are preferred to procedures. The second mechanism is the test for resource compatibility which makes sure that concurrent PoBs do not access non-sharable resources simultaneously.

In [MPT95], we presented a model that takes into account dynamic changes of priorities for PoBs over time to support flexible commitment decisions. This approach is based on the concept of a *degree of satisfaction* for a PoB used by Hanks and Haddawy [HH94] for a decision-theoretic rating of goals, and which is very similar to the concept of dynamic behavioral tendencies presented in the dynamic theory of action in Section 2.2.2. However, it is not yet clear how expensive it can be to determine the dynamic priority of a goal or PoB, and thus, in how far it compromises the requirement of reactivity, and in what cases good heuristics are available for computing the degree of satisfaction. Therefore, in this book we will not discuss dynamic priorities in more detail and refer to [MPT95], instead.

Not only committing, but also *uncommitting* (see [KG91]) has to do with reactivity. At the behavior-based layer, uncommitting means ending the execution of PoBs and inactivating them. Breaking the commitment to a PoB implies being able to recognize the situations corresponding to the exceptions defined for the PoB. Thus, at the behavior-based layer of INTERRAP, the uncommitting task is solved by defining and monitoring the exceptions of active PoBs in the control cycle.

The role of execution granularity. In [RG95], Rao and Georgeff observed that the ability of an agent to react is bounded from below by the time needed to perform its control cycle, as situations are checked and decisions are made only once per cycle. The assumption that underlies most computational models based on cycles of the form sense–recognize–decide–act is that the rate of environmental change is small compared to the time needed to perform a cycle. In INTERRAP, this is achieved by executing PoBs in steps given by the execution function discussed in Section 3.4.2. Here, it is important to find an appropriate execution granularity. In the current model of INTERRAP this granularity is defined at the level of one call to a world interface primitive (see Section 3.3.2 and Chapter 4 for an example).

The idea formulated by Firby [Fir94] in order to keep the granularity of action execution small is to look upon the world interface primitives as processes that are activated rather than as procedures that are called. While the latter perspective implies that the execution procedure blocks until the end of action execution, in the former case, execution can be considered as the (brief) activation of a continuous process; the end of process execution is reported asynchronously by the process itself.

3.4.4 Interfaces

In this section, we shall sum up how the behavior-based layer is embedded into the INTERRAP architecture. This includes three other modules within the architecture: (i) the knowledge-base, (ii) the world interface, and (iii) the local planning layer.

Knowledge base access. The behavior-based layer accesses the lowest part of the knowledge base (see Section 3.3.3), i.e., the world model, where factual information about the environment is stored. Firstly, the basic data structures maintained by the control cycle, such as the descriptions of PoBs and organizatorial information about the PoBs to be monitored, are kept within the agent knowledge base. Secondly, situation recognition, i.e., the checking of activation and monitoring conditions, is performed over the world model using the basic knowledge access functions defined in Section 3.3.3.

Thirdly, in the current version of INTERRAP, belief abstraction has to be done by the control layers; the deductive capabilities of the behavior-based layer are restricted, which is desirable for reasons of reactivity; however, some simple deductive operations can be implemented elegantly as reactor PoBs. E.g., a robot that explores an unknown environment can have a PoB that recognizes neighboring squares of the same type as an area of that type, and that deduces this information automatically.

Connection to the world interface. As we have discussed above in the context of execution granularity, the behavior-based layer calls actoric and sensoric routines defined in the world interface. Apart from that, it needs access to perceptual information like the current sensory values and the content of the message queue. This information is stored in the agent's perception buffer (see Section 3.3.2). By activating the appropriate update routines, the current information is provided in the knowledge base.

Interplay with the local planning layer. This interaction is responsible for achieving the reconciliation of reactive and deliberative behavior in IN-TERRAP. At the beginning of each loop of the layer control cycle illustrated in Figure 3.14, the messages received from the local planning layer (since the last cycle) are looked up and processed. This is done by a method upd_lpl_msg, which is illustrated in Figure 3.15.

The possible messages received from the local planning layer are those defined in Section 3.4.2. The posting of a commitment to executing a procedure has the effect that the corresponding PoB is added to the list of goals; analogously, uncommitting means deleting the PoB from the goal list (line 11ff). The last four commands in Figure 3.15 implement the special functions discussed on page 73f.

The behavior-based layer itself sends the local planning layer an upward activation request in case it has recognized a situation that it cannot cope with. These requests which cause the local planning layer to become active are part of the execution language for PoBs defined on page 71. The language construct call(lpl, do(goal)) is translated by the execution mechanism into a corresponding request message that is sent to the local planning layer. Furthermore, the behavior-based layer sends the local planning layer acknowledgments for each of the requests listed in Figure 3.15.

In summary, the behavior-based layer as the lowest INTERRAP control layer provides the basic reactive behavior of the agent. Furthermore, it implements the procedural knowledge needed to perform routine tasks that the next higher control layer has decided to accomplish. The basic building blocks of the behavior-based layer are patterns of behavior (PoBs). They are maintained in a control cycle, which provides a specific instantiation of the generic

```
1 /*Global variables:                                              */
2   Lpl-calls is list of messages from lpl                         */
3   B-goals, Inact-pob, Int as in the class definition of the BBL */
4
5 meth upd-lpl-msg
6    foreach M ∈ Lpl-calls do
7      if M = commit(B, Args) then
8         B-goals = B-goals + {B};
9         Inact-pob = Inact-pob - {B};
10    fi
11    if M = uncommit(B) then
12       B←uncommit;
13       B-goals = B-goals - {B};
14       Inact-pob = Inact-pob + {B};
15    fi
16    if M = disable(B) then
17       if B ∈ Int then wait-for-termination(B);
18       Inact-pob = Inact-pob - {B} /* remove b from cycle */
19    fi
20    if M = enable(B) then
21       if B ∈ B-goals ∪ Inact-pob then true;
22       else Inact-pob = Inact-pob + {B}; fi
23    fi
24    if M = inhibit(B, Sit) then
25       B←inhibit(Sit);
26    fi
27    if M = permit(B, Sit) then
28       B← permit(B, Sit);
29    fi
```

Fig. 3.15. Method definition for upd-lpl-msgs of the class BBL

functions of an INTERRAP control layer: situation recognition, goal activation, planning, scheduling, and execution.

3.5 The Local Planning Layer

Planning systems are computational systems that, given a description of a goal or a task, produce a sequence of actions the execution of which will lead to the achievement of the goal or task. The development of planning systems is one of the oldest subfields of Artificial Intelligence (see Section 2.2.3). The local planning layer incorporates the facilities of an agent to devise plans for achieving its local goals and for accomplishing its local tasks.

3.5.1 Overview

An architecture for autonomous and interacting agents should support the designer in programming goal-directed agents. It should allow designers of agents to specify actions by preconditions and effects; a planning mechanism should then accept descriptions of instances of planning problems and compute a plan. However, taking into account both the absence of powerful *and* tractable general planning mechanisms (see, e.g., [Byl91] [Byl92]) and our claim that INTERRAP is a general agent architecture, we are faced with a problem: if we commit ourselves to a specific planning mechanism at the local planning layer, this may be adequate for some applications, but inadequate for others. On the other hand, the consequence of making no commitment at all is that the designer of a system must program her own planner for each application.

We suggest a compromise solution for this problem, restricting the possible planning mechanisms from two sides: firstly, any planning mechanism to be used in INTERRAP must provide suitable interfaces to its neighboring layers. As the interface functionality has been kept very general, this is not too severe a restriction. Secondly, the control cycle of the local planning layer determines how the layer basically works. One step within this cycle is the call to a plan generation function, given a description of the planning problem as input. This provides an interface for connecting different (possibly pre-existing) planners.

Thus, the basic idea is to provide an open architectural framework, i.e., a shell into which existing planners can be plugged by the system designer, depending on the domain under consideration. This idea is somewhat idealized, and it has many restrictions. For instance, different planners require different descriptions of the actions available, or they require specific descriptions of the knowledge representing the world states about which they perform reasoning. However, our experience shows that these problems can often be solved in practice by providing appropriate transformation functions (see also the discussion in Section 6.2).

Figure 3.16 overviews the functional structure of the local planning layer. It consists of several functional modules: the controller, the plan generator, the plan interpreter, the plan evaluator, the plan scheduler, and the plan executor. The controller runs the layer control cycle, and it also serves as an interface to the neighboring layers. The plan generator contains the actual planning mechanism. It has access to a plan library with pre-defined plans. The plan interpreter maintains a set of goal stacks. It expands these goal stacks in parallel. The plan evaluator computes the utility of plans, which is used to make a decision on which of the different applicable plans the agent should choose. The plan scheduler creates a schedule from the output of the plan interpreter. A schedule is a temporal partial ordering of plan steps. Finally, the plan executor monitors the schedule and triggers the execution of scheduled tasks when their execution time has come.

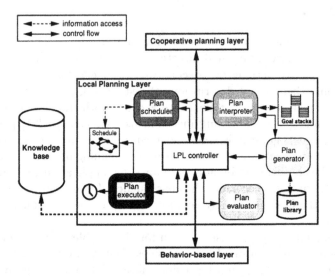

Fig. 3.16. Functional structure of the local planning layer

In this section, the individual components of the local planning layer are described. The layer is instantiated by a specific planning mechanism. In Chapter 4, we show how the approach is used for modeling the loading dock domain. As the primary focus of this work is not on planning itself, but rather on how different important functionalities of an agent (one of which is planning) can be integrated in a control architecture, in this book, we use an approach for planning from second principles for describing the local planning layer. In Section 3.5.2, this approach and the underlying representation of plans, actions, and planning problems are described. Section 3.5.3 outlines the control cycle, and discusses how the five basic functions identified in Section 3.2 are instantiated by the cycle. The section is closed by giving an overview of the role of the local planning layer within the INTERRAP architecture.

3.5.2 Goals, plans, schedules, and plan libraries

In this section, the basic notions and the corresponding data structures are defined that are dealt with at the local planning layer.

Goals. Given a description of an initial state (a situation), a goal describes a set of states plus an attitude the agent has with respect to these states. The states can be characterized either by explicitly listing all the properties that should hold, or they can be described by using a short-hand expression. The latter representation is advantageous if an approach for planning from second principles is chosen (see below). Whenever a plan is stored in a plan library, the corresponding goal states can be encoded in an index; this provides efficient access to plans stored in plan libraries (see [K94]) by means of an indexing function. Efficiently accessing plan libraries is a separate research topic; throughout this work, we assume that an index function exists that allows us to map goal descriptions into plans to their achievement.

The second component of a goal is a modality describing the agent's attitude towards the goal states. The standard attitude, which was used in classical STRIPS planning [FHN71], is the *achievement* of a goal. Recent work on planning investigated further attitudes. In [HH94], Haddawy and Hanks made a distinction between *temporal* and *atemporal* goals. Temporal goals have been further classified into *deadline goals* (achieve the goal by a specified point in time) and *maintenance goals* (keep the goal satisfied during a time interval).

As an other example, in [EHW+92], Etzioni et al. have proposed the language UWL providing three possible attitudes: *satisfy, hands-off,* and *find-out.* Given a formula P, *satisfy P* corresponds to achieving P; *hands-off P* describes the commitment of the planner not to change the truth value of P. This corresponds to the notion of a *maintenance goal* as used by Haddawy and Hanks. Finally, *find-out P* denotes the goal of determining the truth value of P without changing it[6]. The latter type of goal is called *information goal.*

In the planning mechanism defined in this section, we restrict ourselves to the classical notion of achievement goals. Note, however, that the open architecture defined in this book provides the possibility of using a planner providing more complex goal formulae.

Plans. There are two fundamental approaches to planning discussed in the planning literature. The first approach models the generation of plans as a bottom-up process. Starting from a formalization of the preconditions and the effects of actions (operators), planning is understood as searching for a sequence of operator applications in a state space. This is the classical view of *planning from first principles*, which was represented by STRIPS and many other planners (see [AHT90]).

The second approach makes use of the fact that often the search for a plan is not blind, but is guided by domain knowledge. In principle, it is known how specific planning problems in the domain can be solved; the task is to adapt existing plans to specific situations. This perspective of planning is called *planning from second principles.* An example of its use is the notion of scripts defined by Schank and Abelson [SA77]. Moreover, planning from second principles using plan libraries is common in research fields like agent architectures, of which planning is a part, but not the main focus (see [BIP87, BHS93, RG95]).

Using this planning technique allows one to abstract from many problem details in planning. The planner is viewed as a black box, which—given a problem description—returns a plan. However, to be useful for agents acting

---

[6] The difference between a *satisfy* goal and a *find-out* goal is that satisfying that a car is blue might cause the planner to paint it blue, whereas this should not be an option when the goal is to find out that a car is blue [EHW+92, p. 118].

in dynamic environments, such a mechanism has to interleave planning with execution. In this section, we define a mechanism for planning from second principles. The structure of and the access to the plan library are defined. We present a language for representing plans, and discuss how plans are processed.

Plan libraries. A plan library is a collection of plans. It is a list of entries

$$PL = [e_1, \ldots, e_k].$$

The individual entries e_i, $1 \leq i \leq k$ of the plan library are of the form

$$e_i = (\Phi_i, p_i)$$

where Φ_i is the applicability condition of e_i, and p_i is a *plan*. Φ_i is an atomic first-order formula that may contain variables, which are interpreted as universally quantified.

Given a description of a planning problem, how is the access to the plan library defined? The main idea is that a description of a planning problem is transformed into an index formula ψ. In the simplest case, the formula ψ is the description of the goal to be achieved. The operational semantics of plan library access is now defined by a meta function *expand* as

$$expand(\psi, PL) \stackrel{\text{def}}{=} \{p_i \theta | e_i = (\Phi_i, p_i) \in PL \wedge subst.\theta \, with \, \psi = \Phi_i \theta\}.$$

Given an index formula ψ, applying the function *expand* to the plan library PL and ψ returns the set of all plans whose applicability condition matches ψ. One of these alternative plans must be selected for execution (see Section 3.5.3).

This description of the plan library leaves open the structure of plans. As plans for complex tasks cannot be completely specified at planning time, the planning mechanism described in this section is hierarchical. Goals are incrementally refined into subgoals until a sufficiently fine-grained level of abstraction is reached. This plan expansion process is guided by new information that becomes available during plan execution. In INTERRAP, the level of abstraction at which plan expansion stops and plan execution starts is given by the procedure PoBs defined at the behavior-based layer (see Figure 4.4 in Chapter 4 for an example). Given a plan library of the above form, this process of incremental subgoal expansion can be achieved by reading an entry (Φ, p) of the plan library as: "to achieve Φ, expand it to p". Thus, the plan library contains two types of information: firstly, it provides entry points into a hierarchical plan by defining top-level goals. Secondly, it provides information about how abstract plans for achieving abstract goal descriptions can be refined by computing more specific plans achieving more specific subgoals.

Thus plans have a hierarchical structure that is embedded in the plan library. Planning is a recursive process of plan expansion which continues until a level of plan representation has been reached which is suitable for execution. This recursion is embedded in the control cycle (see Section 3.5.3). In the following, we shall look at the representation of plans in more detail.

Plan representation. In this section, we define a language $\mathcal{L}_0$ for the plans stored in the plan library. This language will be used to model the plans for the loading dock domain described in Chapter 4. $\mathcal{L}_0$ consists of four different types of language constructs: primitives, tests, junctors, and control constructs. The primitives are given by the set of procedure PoBs defined at the behavior-based layer. Tests are predicates that are evaluated by sending a query to the agent knowledge base. Junctors are used for composing complex plan expressions from simpler ones. Finally, control constructs are provided to support conditionals and iteration.

Definition 3.5.1 (Plan language $\mathcal{L}_0$). *Let $P = \{p_0, p_1, p_2, \ldots\}$ be a set of primitive plan steps; let $C = \{c_0, c_1, c_2, \ldots\}$ be a set of predicates. Then*

- $\epsilon \in \mathcal{L}_0$ *(empty plan).*
- $p \in \mathcal{L}_0$ *for $p \in P$.*
- *For $p_1 \in \mathcal{L}_0, \ldots, p_n \in \mathcal{L}_0$, we have:*
  - $p_1, \ldots, p_n \in \mathcal{L}_0$ *(sequential composition);*
  - $p_1; \ldots; p_n \in \mathcal{L}_0$ *(disjunctive composition).*
- *For $p_1, p_2 \in \mathcal{L}_0, c \in C$, we have:*
  - **if** c **then** p_1 **else** p_2 **fi** $\in \mathcal{L}_0$*;*
  - **while** c **do** p **od** $\in \mathcal{L}_0$.

Starting from a set of primitive plan steps, $\mathcal{L}_0$ offers the standard junctors *or* (';') and *and* (',') to compose plans. There are two language constructs to direct the flow of control in plan execution depending on the truth value of predicates defined in the knowledge base. Firstly, conditional plans can be built using the **if-then-else** construct. Secondly, iteration can be expressed using the **while-do** construct.

Schedules. Schedules are plan structures that contain temporal information about their execution. The transformation of plan structures into schedules is a first stage of operationalization of plans. Different applications impose different constraints on the design of schedules. These include:

- The ordering of the elements in the schedules, which can be total or partial, leading to linear and nonlinear schedules, respectively.
- The representation of temporal information, which can be either implicit or explicit. In the former case, the ordering relation on the elements of the schedule is interpreted as a precedence relation; in the latter case, there are different possibilities to explicitly represent temporal information, e.g., by specifying starting times for actions, or by specifying time windows defining earliest and latest possible start or finish times (see [FMP96]).

In the following, we adopt a representation, which is based on a total ordering relation and an implicit representation of time. Thus, a schedule is a list $S = [s_1, \ldots, s_n]$, $n \in I\!N$, where $s_i \in \mathcal{L}_0, 1 \leq i \leq n$ are plan steps. The ordering of S is interpreted such that s_i is executed before s_j if $1 \leq i < j \leq n$. To schedule a plan means to insert the plan steps of the plan at an appropriate position within the schedule (see Section 3.5.3).

3.5.3 The control cycle

Plans are selected, interpreted, scheduled, and executed in the *control cycle* of the local planning layer. The definition of the cycle gives the operational semantics of local plans. The cycle is implemented within the LPL controller module (see Figure 3.16). It provides an instantiation of the functionalities of an INTERRAP layer (see Figure 3.7, page 60), which is suitable for dealing with the planning and the execution of local tasks: (1) recognition of local planning situations, (2) generation of planning problems from these situations (this corresponds to goal activation), (3) generation of plans for solving these problems, (4) scheduling of these plans, and (5) their execution.

Overall structure. Figure 3.17 shows the control cycle of the local planning layer. The control cycle is implemented as an object whose main methods (lines 11–16) correspond to the functional boxes shown in Figure 3.16. There

```
/* Definition of the class LPL. The methods process-msg, interpret */
/* are described in detail in Figures 3.18 and 3.19, respectively */
1 class LPL
2 super layer
3   attributes
4      Planstacks      /* list of current planstacks */
5      Msgs     /* inter-layer message queue */
6      Sched      /* current schedule */
7      Plib      /* plan library */
8      [...]
9   methods
10  [...]
11     meth process-msg <?Msg>          /* see Fig. 3.18 */
12     meth generate-plan <?Goal-descr>
13     meth evaluate-plan <?Plan>
14     meth interpret <P-stack>         /* see Fig. 3.19) */
15     meth schedule <Step>
16     meth execute
17  [...]
18
19     meth cycle
20       foreach M ∈ Msgs do
21         process-msg(M); od
22       foreach Ps ∈ Planstacks do
23         if pending(Ps) then true
24         else
25            Step = interpret(Ps);
26            Sched = schedule(Step);
27         fi
28         Sched = execute(Sched);
29       od
```

Fig. 3.17. The control cycle of the local planning layer

are four attributes referencing the data structures that are manipulated in this layer. The attribute `Planstacks` points to a list of plan stacks. These plan stacks operationalize the current intention structure of the agent. Each activation request from the behavior-based layer creates a new plan stack. Thus, multiple goals can be handled. However, this is only possible under the *goal independence assumption*: i.e., we assume that the planner can construct plans for different goals at a time without having to worry about incompatibilities among these goals. The attribute `Msgs` holds the list of messages received from the neighboring layers. The attribute `Plib` provides access to the plan library. Finally, the current schedule of the agent, i.e., the set of actions to which the agent has committed itself but the execution of which has not yet been started, is kept in the attribute `Sched`.

In each loop of the control cycle, first those messages are processed that have arrived from the neighboring layers since the last run of the cycle (lines 20–21). The method `process-msg`, which is used for this purpose, implements the situation recognition and goal activation capabilities of the local planning layer. It is explained in more detail below (see Figure 3.18). In processing these messages, new plan stacks are created and existing ones are removed.

Secondly, the current plan stacks are processed (lines 22–29). In each loop of the cycle and for each plan stack, it is tested whether an action from this plan stack is being executed. If so, the stack is not expanded any further, otherwise one processing step is performed. The processing of the plan stacks includes expanding plans by using the plan library (see page 3.5.2) and interpreting control expressions of the underlying plan language. It is achieved using the function `interpret`, which is described below in more detail. The process of plan interpretation creates new plan steps that are scheduled using the function `schedule` (line 25). Finally, the schedule is monitored for plan steps that are due to be executed.

Situation recognition and goal activation. As most of the actual situation monitoring and the recognition of the external context of situations are achieved by the PoBs at the behavior-based layer (see Section 3.4.3), the implementation of this task at the local planning layer is limited to monitoring for incoming messages from the neighboring layers, which indicate the arrival of a new task or the end of the execution of a procedure PoB. Figure 3.18 shows the definition of the function `process-msg` which implements situation recognition at the local planning layer.

Firstly, messages are classified by their sender. Examples of messages received from the behavior-based layer are upward activation requests (see Section 3.3), the cancelling of a previous request, and the acknowledgment for the execution of a message. Other possible events that are not listed in Figure 3.18, but which are handled similarly, are requests to interpret a given plan or to evaluate a single-agent plan . If an upward activation request is received (lines 4–5), a new plan stack is created, added to the list of plan stacks, and initialized with the top-level goal. The creation of a plan stack implements

```
/* method process-msg of class LPL; ps(id) is the access function */
/* for the plan stack created by the request with id-nr id. Sdr, */
/* id are access functions to the sender and the Id of a message; */

1   meth process-msg(M)
2     if sdr(M) = bbl then
3       case Msg of
4         'do((S, G), Args)':
5             create-planstack((S, G), Args);
6         'stop(Id)':
7             rm-from-schedule(ps(Id), Sched);
8             rm-plan-stack(ps(Id), Planstacks);
9         'done(Id, Stat)':
10            Status = determine-status(Id)
11            if Status = succ then
12                cleanup(ps(Id));
13                else replan(ps(Id)); fi
14            [...] /* similar for other cases */
15        esac
16
17    else
18      case Msg of
19        'commit(Id, P-old, P-new)':
20            popall(Id, P-old);
21          push(Ps, P-new);
22        'uncommit(Id)':
23            rm-from-schedule(ps(Id), Sched);
24            rm-plan-stack(ps(Id), Planstacks);
25        'eval(P)':
26            U = evaluate-plan(P);
27            cpl←inform(done(id(Msg), U))
28        [...] /* similar for other cases */
29      esac
24    fi
```

Fig. 3.18. Processing inter-layer messages in the LPL cycle

the activation of a new goal which is then processed by the plan interpreter
(see Figure 3.17). If a task is cancelled (lines 6–8), the corresponding plan
stack has to be removed, and plan steps that are already scheduled have to
be removed from the schedule.

If a procedure PoB whose execution was triggered by the local planning
layer reports the end of its execution, its success is evaluated by the local
planning layer (line 10). This is done by testing whether the expected effects
of the PoB, which are given by its termination condition TC, correspond to the
current state of the knowledge base. A query kb←query(TC, R, θ) whose
content is the original termination condition is sent to the knowledge base.
The result of the function determine-status is the return value R of the
query.

In some cases, procedures return results that differ from succ or fail. Firby [Fir94] has argued in favor of a more general representation of the outcome of non-atomic tasks, allowing procedures to deliver valuable semantic information about their outcome to the planner. INTERRAP supports this flexible view in two respects: firstly, the planner itself determines whether a PoB has achieved the expected effects by comparing them with the actual effects. Secondly, it is supported in this activity by additional status information from the behavior-based layer. This information may contain any return code; it is stored as the second argument of the done message.

Messages that the local planning layer receives from the cooperative planning layer include the commitment to a plan, the cancelling of a previous commitment, and the request to compute the costs of a plan. Plans to which the cooperative planning layer has committed itself are obtained by projections from multiagent plans describing synchronized courses of action involving more than one agent. A message commit(Stack,Old,New) (lines 19–21), indicating the commitment to a plan New is processed by popping all elements including Old from the corresponding plan stack, and by pushing the new plan on the stack (see also Section 4.3.5 and Figure 4.22 for examples). In case a previous commitment is cancelled, the actions within this plan that are already scheduled are removed from the schedule, and the corresponding plan stack is deleted. Breaking a commitment is necessary if the cooperation process fails during which the plan under consideration was negotiated (see also Section 3.6).

The request to evaluate a plan occurs in different contexts. One context is that the cooperative planning layer needs to determine the cost of the part that the agent plays in a joint plan under negotiation. In this case, an explicit plan structure is sent to the local planning layer together with the request to evaluate it. Another context occurs e.g., if, while running a contract net protocol [DS83] at the cooperative planning layer, the manager of this contract net has to determine the cost of performing the task it is announcing by itself. In this case, the local planning layer receives a request to estimate the costs of performing a task. This in turn may require to devise a plan before evaluating it. The former case is discussed in more detail in Section 3.6. An example for the latter case is the transportation domain; we refer to [FKM94] [FMP95a] for a detailed discussion. Plan evaluation is performed within the local planning layer using the plan evaluator module. It is discussed in the following paragraph.

Plan selection. If the expand function extracts more than one suitable plan from the plan library, one of these candidates has to be selected for execution. As we require that all the plans returned by expand suit the goal description, this selection is *not* critical for the agent's ability to achieve its goal; in principle, a random selection strategy would be sufficient. However, this selection should be based on some notion of *utility*. A plan that achieves a goal with some effort should be preferred to another plan achieving the same goal, but involving a higher effort.

The application of utility theory to planning has been investigated in the area of decision-theoretic planning (see [HRW94] for an overview). However, the focus of decision-theoretic planning has been on using decision theory to make the right decisions when generating plans by selecting those actions that maximize the agent's expected utility. In our case, the utilities of predefined plans are compared to select the plan with the maximal utility.

Definition 3.5.2 (General utility function). *Given a finite set $\mathcal{P}$ of plans, a utility function for plans is a function $u : \mathcal{P} \mapsto \mathbb{R}$, mapping plans into real numbers. The utility of a plan $p \in \mathcal{P}$ is computed as*

$$u(p) = w(p) - c(p),$$

where $w : \mathcal{P} \mapsto \mathbb{R}$ is a worth function, and $c : \mathcal{P} \mapsto \mathbb{R}$ is a cost function for plans.

The worth of a plan according to Definition 3.5.2 is equal to the worth of the goal achieved by successfully executing the plan. As we have to choose among alternative plans for achieving the same goal, we have $w(p_i) = w(p_j)$ for all plan alternatives $p_i, p_j \in \mathcal{P}$, regardless of how the function w is defined. Thus, we can restrict ourselves to comparing the costs of plans. Given a finite set $\mathcal{P}$ of alternative plans and a cost function c, plan selection can therefore be formalized by a function $select : 2^{\mathcal{P}} \mapsto \mathcal{P}$, which returns a plan p with $c(p) = \min_{p_i \in \mathcal{P}} c(p_i)$.

Assigning costs to plans. Assigning costs to plans is straightforward if plans are sequences of atomic actions $\{a_1, \ldots, a_n\}$. In this case, starting from a cost function $\hat{c}$ for atomic actions, the cost of a plan p can be computed by

$$c(p) \stackrel{\text{def}}{=} \sum_{a_i \in p} \hat{c}(a_i).$$

However, evaluating $\mathcal{L}_0$–plan structures is more difficult for two reasons: the first reason are the control expressions contained in $\mathcal{L}_0$. The second reason is the hierarchical nature of $\mathcal{L}_0$ which allows us to express abstract plan steps that are expanded by the plan interpreter at run-time. Thus, plan selection, which is done at planning time, has to be based on estimated costs of plans.

We give an inductive definition of a cost estimation function for $\mathcal{L}_0$–plans. We assume the existence of a cost function for primitive plan steps. The evaluation of conditionals and iterations needs to be based on a probabilistic model allowing to compute probabilities of a condition being true or to predict statistical expected values for how often an iteration has to be carried out.

Definition 3.5.3 (Cost estimation function ($\mathcal{L}_0$)). *Let $\hat{c}$ be a cost function for primitive plan steps. A cost estimation function for $\mathcal{L}_0$–plans is a function $c_0 : \mathcal{L}_0 \mapsto \mathbb{R}$ with:*

- $c_0(p) = \hat{c}(p)$ *for primitive plan steps p.*
- $c_0(p_1, \ldots, p_n) = \sum_{i=1}^{n} c_0(p_i)$ *for $p_i \in \mathcal{L}_0$*
- $c_0(p_1; \ldots; p_n) = \max_{i=1\ldots n} c_0(p_i)$ *for $p_i \in \mathcal{L}_0$.*
- $c_0(\textbf{if } e \textbf{ then } p_1 \textbf{ else } p_2 \textbf{ fi}) = Pr(e) \cdot c_0(p_1) + (1 - Pr(e)) \cdot c_0(p_2)$ *for $p_1, p_2 \in \mathcal{L}_0$. Pr is a probability distribution over conditions; Pr(e) denotes the probability of e being true in a certain world state.*
- $c_0(\textbf{while } e \textbf{ do } p \textbf{ od}) = E(e,p) \cdot c_0(p)$ *for $p \in \mathcal{L}_0$. Here, E(e,p) denotes the statistical expected value for the number of iterations of p required to reach a world state in which e does not hold.*

This cost estimation function provides a rough criterion for choosing among different applicable plans. As it is defined inductively over the structure of plans, it is easy to show that it has the following property: for any two plans $p_1, p_2 \in \mathcal{L}_0$, such that p_1 is syntactically a subplan of p_2 (i.e., each plan step contained in p_1 is also contained in p_2, written $p_1 \sqsubseteq p_2$), we have $c_0(p_1) < c_0(p_2)$.

A difficulty created by the definition of c are the statistical models that are needed to estimate *a priori* the costs of conditional actions and iterations; the construction of these models is nontrivial and requires domain knowledge. Developing statistical domain models is a separate area of research which is closely related to learning; its discussion exceeds the scope of this book, it is also not crucial for the principles developed in this section. We refer to Section 6.2 for a brief discussion.

Plan interpretation. In each control loop, one interpretation step is done for each plan stack, i.e., the agent determines what action to execute next. Plan interpretation is carried out using the function `interpret`. Its definition is shown in Figure 3.19. The elements in the stack that is processed by `interpret` are $\mathcal{L}_0$–plan expressions. Thus, `interpret` defines the operational semantics for plan language $\mathcal{L}_0$.

The plan interpretation function illustrated in Figure 3.19 takes a plan stack as input argument. Its output is the next plan step to be scheduled. While the plan is interpreted, the plan stack is modified accordingly. Each time the function is called, it processes the top element of the stack which is an $\mathcal{L}_0$ plan expression. The **case** control expression determines how the different language constructs of $\mathcal{L}_0$ are processed. In case of a conjunction (lines 6–8), the first element of the conjunction is interpreted by calling the interpretation function recursively, and the remaining elements of the conjunction are pushed onto the stack again for later processing.

A disjunction is handled as follows: the first element of the disjunction is interpreted; the rest of the disjunction is pushed onto the stack again, and marked by a backtrack point '&'. If the execution of the first element returns a success, the backtrack point and the rest of the pending disjunction are removed from the stack (see line 12 in the description of the method

```
/* Input argument Ps denotes a plan stack; the stack operators */
/* empty, push, pop are defined as usual */
1   meth interpret(Ps)
2     if empty(Ps) then return true;
3     else
4         Top = pop(Ps);
5         case Top of
6             'P₁,...,Pₙ':                    /* conjunction */
7                 push(Ps, (P₂, ..., Pₙ)); /
8                 interpret(P₁);
9             'P₁;...;Pₙ':                    /* disjunction */
10                push(Ps, (P₂;...;Pₙ));
11                push(Ps, ('&'));
12                interpret(P₁);
13            'if C then P₁ else P₂ fi':
14                kb←query(C, Retval, θ);
15                if Retval = true then
16                    interpret(P₁θ);
17                    else interpret(P₂θ); fi /* */
18            'while C do P od':
19                kb←query(C, Retval, θ);
20                if Retval = true then
21                    interpret(Pθ);
22                    push(Ps, (while C do P od))
23                else interpret(Ps); fi
24            'beh(Pob, Args)':
25                return bbl←request(activate(Pob, Args));
26            'ctrl-msg(Rcp, Content)':
27                return Rcp← Content
28            default:
29                Newplan = select(expand(Top, Plib));
30                if Newplan = nil then
31                    return cpl←request(do(Top,args(Top)));
32                else
33                    push(Ps, Newplan);
34                    interpret(Ps); fi
35         esac
36    fi
```

Fig. 3.19. The plan interpretation function

process-msg in Figure 3.18). If the execution fails, the corresponding back-track point is retrieved in the plan stack and the next stack element below it, i.e., the disjunction of the remaining alternatives, is interpreted.

The control commands **if-then-else** and **while-do** are interpreted as follows: the conditional branch (lines 13–17) is processed by evaluating the test condition c with respect to the current beliefs of the agent. This is done by sending a query to the agent knowledge base using the query interface. Depending on the result of this query, either the **then** part or the **else** part of the conditional are interpreted. Similarly, the **while** loop (lines 18–23) is interpreted as follows: the test condition is evaluated, if it returns **true**, the body of the loop is interpreted, and the **while** expression is pushed on the stack again.

Lines 24–27 in Figure 3.19 handle two different classes of primitive plan expressions: the one that is most frequently used are steps of the form beh(Pob, Args), which indicate that the plan step is to be interpreted by activating a procedure PoB Pob with arguments Args. The second class of primitive plan expressions comprises all other control messages to the behavior-based layer or to the cooperative planning layer. Examples are the enabling or disabling of PoBs, or upward activation requests to the cooperative planning layer. Note that the primitive plan expressions terminate the recursive calls of the plan interpretation function, since the function call returns the corresponding plan steps. Each call to the function interpret from the control cycle returns exactly one primitive plan step.

Finally, the default case (lines 28–34) deals with plan steps that are neither conjunctions, disjunctions, control expressions, nor primitive expressions. These are expressions denoting abstract plans that have to be expanded using the plan library and the function expand (see page 84). The current plan step is replaced by the result of the plan expansion, and plan interpretation of the expanded plan is called recursively. If no plan is found to deal with the situation, the layer is not competent, and an upward activation request is sent to the cooperative planning layer.

Plan scheduling and plan execution. Actions to which the planner has committed itself have to be scheduled into the agent's intention structure. Depending on the domain under consideration, the complexity of this scheduling task can vary considerably. Therefore, the basic idea is to provide different *scheduling strategies* among which the designer of the planner can choose. In the loading dock domain, scheduling can be accomplished using a very simple scheduling strategy: each action determined by the planning mechanism can be executed *immediately*. This holds true for two reasons: firstly, there are no time constraints specified for actions in the loading dock. Secondly, the plan interpreter interleaves planning with execution in that it expands a plan step only if the previous plan step is finished. Therefore, each plan step determined by the planner can be executed immediately after the corresponding commitment has been made. Moreover, due to our assumption that the goals pursued by different plan stacks are independent, no scheduling overhead is required for plan steps from different stacks.

Other applications require more complex scheduling strategies. For example, in the transportation domain, the output of the planner of a truck agent are routes that are constrained by time windows. For example, one plan step could be *drive from A to B and start between 9* A.M. *and 11.30* A.M. Scheduling this plan step involves recomputing the global schedule by propagating earliest and latest times for departure and arrival through the schedule, and assigning to each action a point in time when the action is actually to be started. In that domain, planning and scheduling are closely intertwined. Furthermore, making the decision when to start executing an action, there is always a trade-off between the risk of inflexibility if commitments are made early and the danger of getting into trouble if commitments are made very late so that unforeseen events cannot be buffered.

Given a scheduling mechanism, the task of the *plan executor* is to ensure that actions are executed in time. This is achieved by monitoring the actions contained in the agent's schedule. When the execution time for one of these actions has come, its execution is triggered by posting the corresponding commitment down to the behavior-based layer.

3.5.4 Interfaces

The local planning layer interacts with three other modules of the INTERRAP agent, i.e., (i) the knowledge base, (ii) the behavior-based layer, and (iii) the cooperative planning layer. In this section, the interface between the local planning layer and these modules is summarized.

Knowledge base access. Like the behavior-based layer, the local planning layer can access the world model layer of the agent knowledge base in order to obtain information about the world. This has been described in detail in Section 3.4.4. Additionally, it has access to the mental model part of the knowledge base where plans, goals, schedules, and plan libraries are stored that are relevant for the local planning layer. In the current implementation of the INTERRAP agent model, the knowledge defining the agent's mental model is kept locally within the control layer and the access to these data structures does not require physical access to the agent knowledge base; thus, so far, the notion of a mental model within the agent knowledge base is conceptual rather than physical.

Interplay with the behavior-based layer. There are two main modes of interaction between the local planning layer and the behavior-based layer: on the one hand, the local planning layer receives upward activation requests by the behavior-based layer; on the other hand, commitments to the execution of procedure PoBs are posted down to the behavior-based layer. There are additional actions, like the enabling, disabling, inhibition, and permission of PoBs by the local planning layer, which have been discussed in Section 3.4.4. Messages that are sent to the behavior-based layer are generated by the planning mechanism. The only exception are acknowledgments of the form (inform, done(...)), which are generated by the control cycle after a request from the behavior-based layer has been carried out. Incoming messages are processed within the control cycle by the function process-msg (see Figure 3.18).

Local and cooperative planning. The interface between the local and the cooperative planning layer defines how local task planning is coordinated with multiagent planning: Firstly, whenever a problem cannot be solved satisfactorily by the local planning layer, an upward activation request to the cooperative planning layer is generated. This is shown in the definition of function `interpret` (Figure 3.19, line 26). Secondly, the local planning layer receives commitments to the execution of single-agent plans from the cooperative planning layer as a result of the cooperative planning process; the single agent plans are projections of the activities of the agent in a *joint plan*. The representation and generation of joint plans and their translation into single-agent plans shall be explained in detail in Section 3.6.

Another important link between local and cooperative planning is provided by the ability of the local planning layer to evaluate single-agent plans based on a local utility function. This functionality can be accessed by the cooperative planning layer by using the message `eval(Plan)` to evaluate the local utility of a joint plan (which may be proposed by another agent) for the agent.

Finally, we shall briefly discuss the abovementioned notion of not being able to solve a problem *satisfactorily*, which has been used as a criterion for triggering cooperative planning from the local planning layer. There are at least two different perspectives of this criterion: the first perspective is that cooperation should be triggered if a problem *cannot be solved* by local planning. E.g., in the loading dock, a conflict situation where two forklifts block each other in a narrow shelf corridor, requires synchronization, and cannot be resolved by an individual agent. Other tasks cannot be performed by an individual agent due to lacking resources or due to lacking skills. For example, a truck may not be able to carry out an order *in time*. The second perspective is that a task may be done locally, but that *better solutions* can be achieved by cooperation. Actually, this perspective is not part of the modeling of the loading dock domain provided in Chapter 4; however, it is inherent in the transportation domain: a shipping company which receives a transportation order is in most cases able to carry it out by using its local resources, i.e., the set of its own trucks. Often, however, doing so involves high costs, and selling the order to another company may be beneficial for the company. Thus, one way of implementing the notion of solving a problem satisfactory is by computing utilities for solutions and by triggering a cooperation in case the local utility falls short of a certain threshold. However, such an extension is subject to future work. Possibilities of its integration are discussed in Section 6.2.

In summary, the local planning layer of an INTERRAP agent incorporates the agent's capabilities of planning to achieve its local goals. We have introduced a mechanism for planning from second principles, which makes use of precompiled, hierarchical plans stored in a plan library. The primitive plan steps about which the planner performs reasoning are procedure PoBs defined in the behavior-based layer. The local planning layer interacts with the cooperative planning layer to cope with situations requiring multiagent planning.

3.6 The Cooperative Planning Layer

Autonomous agents that co–exist in multiagent environments must coordinate their activities with those of other agents. An important tool for coordination is negotiation; the basic precondition for negotiation is communication. The cooperative planning layer of the INTERRAP agent supports general negotiation mechanisms allowing agents to agree on common goals, to allocate tasks, to resolve conflicts, or to coordinate their local plans. It is activated by the local planning layer to cope with situations that cannot be dealt with satisfactorily by local planning as they require taking into account knowledge about other agents or as they require explicit synchronization.

3.6.1 Overview

The cooperative planning layer provides functions for recognizing interaction situations, and for deriving goals from recognized situations. It also contains tools to select protocols and strategies for negotiation, and to provide agreements on certain negotiation topics with other agents. Throughout this work, we assume that agents negotiate about **joint plans** that allow them to satisfy a set of individual goals (i.e., to resolve goal conflicts cooperatively), or to satisfy a common global goal. The structure of the cooperative planning layer is illustrated in Figure 3.20. It runs in a control cycle where requests from the local planning layer are received, additional information is gathered, and then, a suitable protocol and a negotiation strategy is selected. The planning, scheduling, and execution tasks of the cooperative planning layer include the processing of protocols using negotiation strategies. In the course of protocol execution, the agent must collect enough information about other agents' goals to be able to classify the interaction situation and to generate the input description of a cooperative planning problem. Informally, this input description consists of a situation description which is a part of the agent's world model, of a goal of the agent, and of a set of goals, one for any other agent involved in the interaction. The solution for such a multiagent planning problem is a suitable joint plan.

Planning, scheduling, and execution take place at two different levels in the cooperative planning layer: at the meta-level, protocols are processed according to strategies; at the object-level, plan-generating routines are called from within protocols in order to generate the negotiation set [RZ94], i.e., the set of candidate plans.

The actual *multiagent planning problem*, i.e., the problem of finding a suitable sequence of actions resulting in a state that satisfies the union of

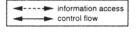

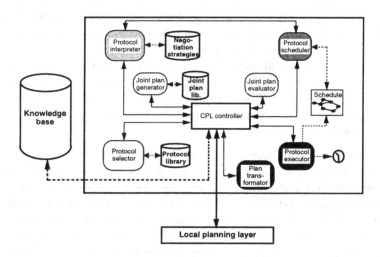

Fig. 3.20. The cooperative planning layer

the goals of the agents involved, is addressed at the object-level. The *plan agreement problem*, i.e., the problem of how a set of rational autonomous agents can agree on one from a set of candidate plans, is solved at the meta-level by negotiation protocols.

This section is structured as follows: in Section 3.6.2 the notions of negotiation protocols and strategies are introduced and the underlying model of negotiation is presented. Section 3.6.3 then explains the concept of joint plans and their representation. In Section 3.6.4, the structure of the control cycle is described and the individual phases of evaluation, agreement, transformation, and execution of joint plans are discussed. The interfaces to neighboring layers are discussed in Section 3.6.5.

3.6.2 Negotiation

Negotiation (see e.g., [Syc87] [CKLM91] [RZ94]) is a mechanism allowing autonomous agents to find mutual agreement on a matter. Negotiation is used to resolve conflicts ([Syc87], [Kle91]), or to allocate tasks among agents ([DS83], [CKLM91]). In this section, we describe a negotiation mechanism, which we adopted from that presented by Rosenschein and Zlotkin in [RZ94].

The general negotiation framework. In [RZ94, p.35ff], three main issues are described that determine coordination among agents:

1. The **space of possible solutions** or *deals*; the set of possible deals among which the agents have to find an agreement is called *negotiation set*.

2. The **negotiation protocol**; given the negotiation set, how can agents proceed to converge towards an agreement? Negotiation protocols restrict the possible courses of negotiation by specifying actions and possible reactions in the negotiation process.

3. The **negotiation strategy**; given a negotiation set and a protocol, the negotiation strategy accounts for the decisions of an individual agent. The strategy is necessary because protocols do not fully determine the course of a negotiation, but mark branching points at which the agent has to make decisions.

The authors used this model as a starting point for the investigation of properties of and interrelationships between negotiation protocols and strategies. However, embedding a negotiation framework based on this model into a practical agent architecture is difficult for several reasons: firstly, in a dynamic agent environment, even determining the topic of negotiation is a difficult task[7]; even determining the agents participating in the negotiation can be a problem. For example, one agent may recognize a certain conflict while the other does not. On the other hand, in order to be able to decide what type of negotiation should be started, agents may need additional information.

Secondly, Rosenschein and Zlotkin assume that the negotiation set is fully determined and *known to all participants* before the actual negotiation starts. Apart from the fact that this form of common knowledge may not be achievable under certain conditions [HM90], the computation of the negotiation set may be prohibitively expensive (e.g., computing all possible joint plans) and may be subject to negotiation itself. For example, Durfee and Lesser [DL89] have proposed to achieve negotiation on a joint plan by announcing partial plans and designing the global plan incrementally.

In order to overcome these problems, we suggest the following procedure, which is summarized in Figure 3.21: firstly, the negotiation process is preceded by an information gathering phase to determine whether negotiation is required/possible at all, and, if so, to determine the topic of negotiation. Secondly, the negotiation set is generated. Assuming safe communication, common knowledge[8] can be achieved by uniquely determining one agent to compute the negotiation set (see below), and by having this agent broadcast it to all other agents. Thirdly, agents must agree on a negotiation protocol. Fourthly, roles are assigned to the agents in the negotiation, corresponding to the different parts they play in the protocols. If protocols are symmetrical, i.e., the role assignment is not constrained by the nature of the protocol,

---

[7] Rosenschein and Zlotkin assume that the description of the problem is clear to all agents, and that the negotiation set can be computed based on this description.

[8] At this stage, the weaker notion of *everybody knowing that* can be sufficient. See [FHMV95] for a discussion of *common knowledge* vs. *everybody knows*.

agents have to agree on a role assignment. In asymmetrical protocols, the role assignment is determined in advance. An example of an asymmetrical protocol is the contract net protocol, where, for a given subject, it is clear who is the manager and who are the bidders. In order to avoid infinitely nested negotiation, we assume that agents "toss a coin" in order to determine the roles in the negotiation in the symmetrical case (e.g., by using the election algorithm defined in Figure 3.22). Finally, each agent chooses its local negotiation strategy; then, the negotiation starts. In the remainder of this section, we describe how the individual steps of this process are modeled in INTERRAP.

```
1.  Exchange status or goal information among the agents.
2.  Determine negotiation topic. If no negotiation
    required/possible, then goto 7; otherwise continue.
3.  Determine negotiation protocol and assign roles to agents.
4.  Elect leader to compute the negotiation set.
5.  Broadcast negotiation set to all agents.
6.  Negotiate; if solution agreed then exit negotiation;
    otherwise continue.
7.  Abort negotiation.
```

Fig. 3.21. The negotiation process

Information gathering and fixing the topic. The first two phases of the negotiation process shown in Figure 3.21 determine the problem to be solved, i.e., the multiagent planning problem consisting of the initial situation and the goals of the participants. The individual tasks that have to be performed in this process include finding out whether the other agent is interested in or capable of negotiation[9] and determining the goals of the agents involved.

There are two basic mechanisms for obtaining information about other agents' goals: either goal information may be communicated, or it may be observed using appropriate goal recognition tools [Tam95]. Throughout this book we assume that agents are willing to reveal their goals truthfully to other agents if asked, and that communication is safe. These two assumptions guarantee that each agent will achieve the necessary information about others' goals after finite time. For a discussion of negotiation in non-cooperative or adversarial domains, we refer to [RZ94, Chapters 4 and 6]. Given that each agent has the necessary information about the topic, we further assume that it is common knowledge which negotiation protocol to choose.

Election protocols. Given a negotiation topic and a protocol, the different roles in the protocol have to be distributed among the individual agents.

---

[9] An INTERRAP agent must not necessarily have a cooperative planning layer (see Section 4.4); negotiation with such an agent is not possible.

For example, in a contract net protocol, it has to be determined who is the manager and who are the bidders; in a conflict resolution problem among different agents, one of the agents is supposed to generate the negotiation set, i.e., the set of candidate joint plans that may be used to resolve the conflict. Whereas role assignment is deterministic in some interactions, it requires a separate process of coordination in other cases. For example, in a blocking conflict between two robots in the loading dock, each of these robots is capable of generating the negotiation set. Such situations are called symmetrical.

The underlying general problem is to select one representative from a set of otherwise equal agents; this problem is called the **election problem**. The election procedure is not required to be fair. While we could look at this problem as a negotiation problem itself, we will not do so in this work. What we are interested in are mechanisms allowing agents to resolve conflicts by cooperative planning or to allocate tasks among each other. Therefore, we are looking for a simple and efficient way of solving the problem of finding role assignments and of determining which agent should compute the negotiation set.

Distributed operating systems theory [Mat89] offers so-called *election algorithms* for solving this class of problems. In election algorithms, agents have unique identification numbers (ID); as we do not require fairness of the election, it is sufficient to solve the problem by ensuring that the agent with the highest ID wins. The election problem can be mapped to the graph-theoretic problem of having each agent in the net know the number of the agent with the highest ID. In a graph G with e edges, n nodes, and k initiators of the election process, the average message complexity of the election problem is $\mathcal{O}(e + n \log k)$ [Mat89].

Figure 3.22 shows an algorithm for solving the election problem, assuming the problem of global termination being solved (each agent knows whether it knows the highest ID in the net)[10]. Note that each agent participating in the election is required to use this algorithm. The algorithm is an adaptation of the Chang-Roberts election algorithm [CR79]. It allows a (possibly homogeneous) group of agents to agree on a leader; this protocol can be used to solve the role assignment problem in symmetrical cases, and to elect one agent to generate the negotiation set.

Generation of the negotiation set. The agent that won the election generates the set of possible solutions for the negotiation. How this generation is done, and whether it can be performed by an individual agent at all, depends on the negotiation domain. In some cases, it is very easy to determine

---

[10] The termination problem in distributed systems is very hard in general (see [Mat89, Chapter 4]). However, it can be easily solved if some simplifying assumptions are made. In our context, we assume that each agent knows the number of other agents participating in the election. In that case, an agent knows that it knows the highest ID if it has received a message from each other agent.

```
      meth election(I)
      { M = I; /* I denotes the ID of the agent */
        send I to all neighbors
        on reception of message J do
          if M < J then
            M = J;
            send J to other neighbors /* if any */
          fi;
        on message from each neighbor received do
          if M = I then return leader else return follower fi
```

Fig. 3.22. An election algorithm

the negotiation set: for example, in a bilateral seller-buyer negotiation, it is normally defined by the interval $[o_b, o_s]$ where o_b is the first offer of the buyer and o_s is the first offer of the seller. In a contract-net–like auction process, the manager may define the negotiation set by prescribing a range for admissible bids for a given task. In extended forms of the contract net where task decomposition is part of the protocol itself (see e.g., [FMP95a]), the negotiation set may either consist of a set of possible task decompositions with associated costs, or it may emerge as a consequence of the bidding process. The latter procedure makes sense in case the number of possible bids prohibits their explicit enumeration, e.g., in distributed scheduling problems.

Our negotiation domain is cooperative planning and the agreement on joint plans to resolve given conflict. The number of plans can be exponential in terms of the number of agents and of the actions they may perform; this renders the task of generating the whole set of plans impossible. However, in the domain we are looking at, the number of reasonable *abstract* plans for typical conflict situations is small. Abstraction means that the plans that are proposed are not fully instantiated and are described by abstract plan steps, which is supported by the representation of plans in INTERRAP (see Section 3.5.2). Thus, we assume that each conflict situation can be mapped into a small set of "reasonable" plans that can be extracted from a plan library. This procedure shall be described in more detail in Section 3.6.3.

The negotiation. At this stage, we know the topic of the negotiation, the agents participating in the negotiation, the negotiation set, the protocol to be used, and the roles of the individual agents. We now formally define the negotiation process.

Definition 3.6.1. *Let D be a negotiation domain; a negotiation is a tuple*

$$NEG = (A, R, \rho, N, U, P, S)$$

where

- $A = \{a_1, \ldots, a_k\}$, $k \geq 2$, is a set of agents a_i with mental states B_i; B_i consists of the informational, motivational, and deliberative state of a_i according to Section 3.2.
- $R = \{r_1, \ldots, r_l\}$, $l \leq k$ is a set of roles;
- ρ is a function that assigns roles to agents: $\rho(a) = r$ for $a \in A$, $r \in R$.
- $\emptyset \neq N \subseteq D$ is the negotiation set.
- $U = \{u_1, \ldots, u_k\}$, where $u_i : N \mapsto \mathbb{R}$ is the utility function for agent a_i.
- $P = (K, \{\pi : R \times K \mapsto 2^K | r \in R\})$ is a negotiation protocol. Here, $K = \{k_1, \ldots, k_m\}$ is a finite set of communication primitives. There are two distinguished primitives $\{start, done\} \subseteq K$. They facilitate internal control and are not communicated. π maps communication primitives into a subset of admissible reactions with respect to a specific role r within the protocol.
- $S = \{\sigma_i : P \times R \times K \times 2^D \times U \mapsto K \times N | 1 \leq i \leq k\}$ is a set of negotiation strategies, one for each agent. The input of σ_i is the current protocol and the role of the agent in the protocol, a received (possibly parameterized) message, the current negotiation set, the agent's utility function, and its current internal state. The output of σ_i is a reaction (i.e., the reply to the received message), the modified negotiation set and internal state. In the following, the internal state B_i is not explicitly represented as an input/output argument of σ, i.e., $\sigma_i(P, r_i, k, N, u_i) = (k', N')$ with $k' \in \pi(r_i, k), N' \subseteq N$. In case agent a_i determines the protocol and the role, the input arguments k and r can be omitted.

While the mental state B_i of agent a_i is not explicitly represented in the definition of negotiation strategies, it is considered an input/output argument of σ. I.e., a strategy both makes use of the mental state of the agent and may modify it.

In the following, the notions *negotiation protocol* and *negotiation strategy* are explained in more detail.

Negotiation protocols. specify the set of possible reactions to a message that has been received from another agent. As protocols are described in terms of roles, these reactions are role-specific. A protocol is defined by a set of message types characterizing the admissible messages, and by a set of functions restricting the admissible courses of the communication process.

Message types. The idea of semantically classifying the set of messages that are exchanged among agents by a finite set of message types has a long tradition in AI and multiagent systems (see [CP86] [FF94]). For the examples throughout this work, we use the following message types: INFORM, ACCEPT, ANNOUNCE, PROPOSE, REJECT, REQUEST, GRANT, MODIFY, CONFIRM. The general question of the existence of a "sufficient" finite set of speech act types has been discussed in the literature; we refer to [Lux95] for a summary of this discussion.

Representation of protocols. Given a set K of cooperation primitives, a protocol is defined by the set of possible reactions to each primitive in K for each role. This is done by role-specific functions π. At a first glance, this is similar to nondeterministic finite automata (NDFA); however, as we shall see later on, the representation of the negotiation process as an NDFA is too restrictive to be of practical use: certain decisions in negotiation rely on information about the current internal state of the agents, and certain protocols depend on information about the current negotiation process. We incorporate this context-dependent component into our model by taking into account the mental state of the agent in its negotiation strategy.

Use and generation of protocols. There are strong parallels between negotiation protocols and joint plans: whereas plans restrict and synchronize the activities of agents in the world, protocols restrict and synchronize the communication process itself; thus, negotiation protocols can be looked upon as *meta joint plans* whose result is a decision, e.g., about what plan to use. Thus, as is the case with joint plans themselves, there are two ways of using protocols. One is to regard them as given conventions that are known to all participants of a negotiation. The second way of handling negotiation protocols is by generating them upon demand.

However, only little research has been done so far towards an automation of these meta-level coordination processes[11]. One reason for this is that there is no formal semantics that is generally agreed upon for the basic building blocks of communication protocols, i.e., for the message types introduced above. Even advanced languages such as KQML [FF94] that are using these message types (called *performatives* in KQML) do not yet provide such a semantics [MLF96].

In our approach, the semantics of message types (speech acts) is given operationally by the protocol functions π and σ providing possible reactions of the recipients of these messages. This reveals a further problem in defining a general semantics for message types, as it not only depends on the message type itself, but also on its content; for example, whereas the protocol function π can restrict the possible reaction to a message PROPOSE p, $p \in N$, as being either ACCEPT or REJECT, the decision which of both reactions is appropriate mainly depends on the agent's attitude towards content p. This, however, may depend on the mental state of the agent, i.e., on its beliefs. It is this context-dependent component of agent decision-making that is expressed by the negotiation strategy σ explained in the following paragraph.

Negotiation strategies. While negotiation protocols help restrict the possible courses of negotiation, they do not provide unique and nondeterministic descriptions of possible negotiations. Rather, in each step of the negotiation, an agent may have to choose among different possible ways of how to continue, i.e., it follows its own strategy in a negotiation; even if two agents play identical roles in a negotiation, their strategies can be different (e.g., two bidders

---

[11] The work of Shoham and Tennenholtz [ST92] on social laws is a notable exception.

in a contract-net like auction protocol). We define strategies for agents relative to protocols and roles; in doing so, we assume that agents do not take information into account about earlier negotiations[12]. However, as opposed to [RZ94], we allow that the decisions of an agent in a specific situation in a protocol depend not only on the previous action of the partner/opponent in the negotiation, but also on additional information about the current negotiation process. This requirement is necessary to be able to model e.g., the contract net protocol (see below).

Example: joint plan negotiation. In the first example, assume that two agents a_1 and a_2 who have run into a conflict have to agree on a joint plan. The negotiation set has been computed to consist of four plans, namely $\{p_1, p_2, p_3, p_4\}$. There are two roles in the protocol: $R = \{leader, follower\}$. Let the role assignment function ρ be a random function with $Pr(\rho(a_1) = leader) = Pr(\rho(a_2) = leader) = 0.5$. Let the protocol be $P = (K, pi_{JPN})$, where $K = \{\text{PROPOSE}, \text{ACCEPT}, \text{MODIFY}, \text{CONFIRM}\}$. Let p, p', p'' be plans, and let π_{JPN} define the protocol as:

$\pi_{JPN}(start) = \{\text{PROPOSE}(p)\}$
$\pi_{JPN}(\text{ACCEPT}(p)) = \pi_{JPN}(\text{ACCEPT}(p)) = \{\text{CONFIRM}(p)\}$
$\pi_{JPN}(\text{MODIFY}(p, p')) = \pi_{JPN}(\text{MODIFY}(p, p')) = \{\text{ACCEPT}(p'), \text{MODIFY}(p', p'')\}$
$\pi_{JPN}(\text{PROPOSE}(p)) = \{\text{ACCEPT}(p), \text{MODIFY}(p, p')\}$
$\pi_{JPN}(\text{CONFIRM}(p)) = \pi_{JPN}(\text{CONFIRM}(p)) = \{done\}$.

Beginning with a state *start*, the leader proposes a solution from the negotiation set. The follower either accepts the proposed solution or it makes a counterproposal; the latter is indicated by the MODIFY message. This process continues until either agent accepts the other's proposal. In that case, the proposing agent confirms the deal, and the protocol is finished. Note that the joint plan negotiation protocol can be represented as an NDFA, as the reaction of each agent only depends on the previous action of the negotiation partner.

Figure 3.23 shows the graphical illustration of P. Assume that agent a_1 in the above example has been elected leader. According to the protocol π_{JPN}, it starts by proposing a plan from the negotiation set. But which plan should a_1 choose? And, assuming a_1 chooses plan p_1, should agent a_2 accept this plan or should it make a counterproposal? These questions are answered by the strategies of the agents, given their local utility functions u_1 and u_2. In the example, the strategies σ_1 and σ_2 for a_1 and a_2 are defined as follows[13]:

$\sigma_1(start, N, u_1) =$

---

[12] This was called the *isolated encounter assumption* by Rosenschein and Zlotkin.
[13] As there are only two agents, we unify the strategy for a_1 with the leader strategy, and the strategy of a_2 with the follower strategy. We also omit the name of the protocol as the first argument of the strategy. If the strategies for both agents are equal, we denote this by the subscript i in σ_i.

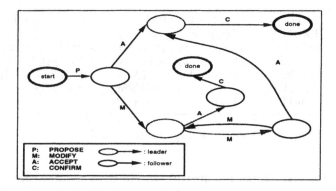

Fig. 3.23. A protocol for joint plan negotiation

$\{(\text{PROPOSE}(p), N) \text{ with } p \in N \text{ with } u_1(p) = \max_{p' \in N} u_1(p')\}$

$\sigma_i(\text{PROPOSE}(p), N, u_i) =$
$\quad \{if \ u_i(p) \geq \max_{p' \in N} u_i(p') then \ (\text{ACCEPT}(p), N)$
$\quad\quad else \ (\text{MODIFY}(p, p'), N - \{p\}) \text{ with } p' \in N, u_i(p') = \max_{p' \in N} u_1(p')\}$

$\sigma_i(\text{ACCEPT}(p), N, u_i) = (\text{CONFIRM}(p), \{p\})$

$\sigma_i(\text{MODIFY}(p, p'), N, u_i) =$
$\quad \{N' = N - \{p\};$
$\quad if \ u_i(p') \geq \max_{\hat{p} \in N'} u_i(\hat{p}) then \ (\text{ACCEPT}(p'), \{p'\})$
$\quad\quad else \ (\text{MODIFY}(p', \hat{p}), N' - \{p'\}) \text{ where } \hat{p} \in N' \text{ with } u_i(\hat{p}) = \max_{\hat{p} \in N} u_i(\hat{p})\}$

$\sigma_i(\text{CONFIRM}(p), N, u_i) = (done, N)$

This protocol is similar to the monotonic concession protocol that was presented in [RZ94]. Each agent starts with the offer that maximizes its utility. If the other agent rejects an offer, it proposes a counteroffer. In case an offer is rejected, the agent who made this offer deletes the corresponding element from the negotiation set. Similarly, if an agent rejects an offer, it deletes this element from the negotiation set. The following theorem states the termination of the protocol:

Theorem 3.6.1. *Let N be a finite negotiation set, and let u_i induce a total ordering on N. Then, a negotiation which is carried through using $K, \pi_{JPN}, \sigma_1,$ and σ_2 terminates after a finite number of steps.*

Let us put this more clearly by tracing the above example. Assume that the utility functions for the plans in the negotiation set are given by:

$u_1(p_1) = 4, u_1(p_2) = 3, u_1(p_3) = 2, u_1(p_4) = 1$
$u_2(p_1) = 1, u_2(p_2) = 2, u_2(p_3) = 3, u_2(p_4) = 4.$

Figure 3.24 shows a trace of the negotiation process, during which a_1 and a_2 agree on plan p_3.

a_1:	$(start, \{p_1, p_2, p_3, p_4\}, u_1) \rightarrow (\text{PROPOSE}(p_1), \{p_1, p_2, p_3, p_4\})$
a_2:	$(\text{PROPOSE}(p_1), \{p_4, p_3, p_2, p_1\}, u_2) \rightarrow (\text{MODIFY}(p_1, p_4), \{p_4, p_3, p_2\})$
a_1:	$(\text{MODIFY}(p_1, p_4), \{p_1, p_2, p_3, p_4\}, u_1) \rightarrow (\text{MODIFY}(p_4, p_2), \{p_2, p_3\})$
a_2:	$(\text{MODIFY}(p_4, p_2), \{p_4, p_3, p_2\}, u_2) \rightarrow (\text{MODIFY}(p_2, p_3), \{p_3\})$
a_1:	$(\text{MODIFY}(p_2, p_3), \{p_2, p_3\}, u_1) \rightarrow \{N = \{p_3\}; (\text{ACCEPT}(p_3), \{p_3\})\}$
a_2:	$(\text{ACCEPT}(p_3), \{p_3\}, u_2) \rightarrow \{(\text{CONFIRM}(p_3), \{p_3\}); done\} \}$
a_1:	$(\text{CONFIRM}(p_3), \{p_3\}, u_1) \rightarrow (done, \{p_3\})$

Fig. 3.24. A trace of the joint plan negotiation example

Example: the contract net protocol. The contract net protocol [Smi80, DS83] is *the* classical protocol for task allocation in MAS. Its original purpose has been to allocate a task among a group of distributed problem solvers. It has been adopted and further developed by different researchers for a variety of distributed resource management problems. In the contract net, a *manager* agent who has a task to be performed is looking for the most suitable from a set of *bidder* agents. For instance, imagine that you are the editor of a journal who has a bunch of papers to be reviewed by a group of busy experts. Allocating papers to experts is a matter of qualification, but editing a journal is also a matter of time and of keeping to deadlines. Whereas you know perfectly well who might be expert for a specific category of papers, you do not know which one of your potential experts can do the review quickly. Thus, you are the manager in the contract net, and you send an order for reviewing a paper to each expert; the experts return the estimated costs (i.e., how quickly they can do the review). Assuming that the bidders are truthful, you grant the review to the expert who can do the job fastest, and reject all others. The protocol ends with the winning bidder reporting the (hopefully) successful completion of its task.

In the negotiation model presented so far, the contract net is formalized as follows: the set of agents is $A = \{a_1, \ldots, a_k\}$. The set of roles is $R = \{manager, bidder\}$. The role assignment function ρ is given by $\rho(a_1) = manager$, $\rho(a_2) = \ldots = \rho(a_k) = bidder$. The protocol is $P = (K, \pi_{CNP})$, where $K = \{$ ANNOUNCE, INFORM, GRANT, REJECT, REPORT$\}$, and π_{CNP} describes the subprotocols for manager and bidder. In the following, let t be a task, let v denote a value, and, for a task t, let *status(t)* be a function denoting the status of the execution of t:

$\pi_{CNP}(start) = \{\text{ANNOUNCE}(t)\}$
$\pi_{CNP}(\text{ANNOUNCE}(t)) = \{\text{INFORM}(t, bid(v))\}$
$\pi_{CNP}(\text{INFORM}(t, bid(v)) = \{\text{GRANT}(t, bid(v)), \text{REJECT}(t, bid(v))\}$
$\pi_{CNP}(\text{GRANT}(t, bid(v))) = \{\text{REPORT}(t, status(t))\}$
$\pi_{CNP}(\text{REJECT}(t, bid(v))) = \{done\}$
$\pi_{CNP}(\text{REPORT}(t, status(t))\}) = \{done\}$

Figure 3.25 graphically illustrates the contract net protocol. The negotiation set describes the current interval of possible solutions; it is computed

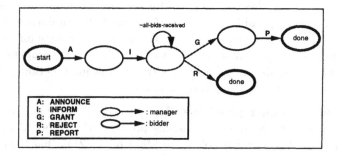

Fig. 3.25. The contract net protocol

incrementally: if not restricted otherwise, e.g., by internal constraints of the manager, the negotiation set in the beginning of the contract net is $[-\infty, 0]$, specifying the least and highest expected utility for the manager which is computed by the costs specified in the bids.

In contrast to the joint plan negotiation protocol described before, the contract net is a multilateral negotiation protocol. Another important difference between the two types of negotiation is that the contract net is an example of a negotiation process that cannot be modeled as an NDFA, as it requires the manager to keep track of the number of bids received to determine when to make a decision as to grant or reject an order. This is reflected in the following definition of the negotiation strategies σ_m and σ_b for the manager and the bidder:

$\sigma_m(start(t), N, u_m) =$
 $\{(\text{ANNOUNCE}(t), [-\infty, 0]) \text{ to all bidders }\}$
$\sigma_b(\text{ANNOUNCE}(t), N, u_b) =$
 $\{\text{if } can\_do(t, u_b(t)) \text{then}$
  $(\text{INFORM}(t, u_b(t)), [u_b(t), 0])$
  $\text{else } (\text{INFORM}(t, \infty), N)\}.$
$\sigma_m(\text{INFORM}(t, bid(v), N, u_m) =$
 $\{$ $bids\_received = bids\_received + 1;$
  $\text{if } v > min(n \in N) \text{ then } N = [v, 0];$ /* note: v < 0 */
  $\text{if } bids\_received = k - 1 \text{ then}$
   $\text{send } (\text{GRANT}(t, v), \{v\}) \text{ to bidder with } bid(t, v) = \max;$
   $\text{send } (\text{REJECT}(t, v), N) \text{ to all other bidders; }\}$
$\sigma_b(\text{GRANT}(t, v), N, u_b) = (\text{REPORT}(t, status(t)), N)\})$
$\sigma_b(\text{REJECT}(t, v)), N, u_b) = (done, N)$
$\sigma_m(\text{REPORT}(t, status(t)), N, u_m) = (done, N)$

3.6.3 Joint plans

An important means for coordination among multiple agents are *joint plans*[14], i.e., plans that represent both actions of multiple agents and coordination

---

[14] We use the terms *joint plan*, *multiagent plan*, and *cooperative plan* interchangeably.

relationships among these actions. In this section, we define the notion of a multiagent planning problem, which is the input to the problem-solving process at the cooperative planning layer. We define a language for joint plans and describe the representation of *joint plan libraries*. The operational semantics of a joint plan is defined by a transformation function mapping it into a set of synchronized single-agent plans.

Multiagent planning problems. The purpose of the cooperative planning layer is to solve multiagent planning problems (MAPPs). MAPPs differ from single-agent planning problems (see Section 3.5.2) in that the goal state is described as a set of single-goals each of which is labeled with the corresponding agent. Often, these goals are in conflict (we say that they *block* each other), and to solve the MAPP means to resolve this conflict. As in certain domains agents can be individual rational and self-interested, and as they are resource-bounded and do not have complete information about the state of the world, a global description of MAPPs is not useful. Rather, MAPPs are defined from the perspective of an individual agent: the description of the initial situation is given by a portion of the agent's world model, and agents are able to distinguish between their own goal and between that of the other agents involved in the MAPP.

Definition 3.6.2 (Multiagent planning problem). *Let $A = \{a_1, \ldots, a_n\}$, $n \in \mathbb{N}$, be a set of agents. For $a_i \in A$, let S_i be a subset of a_i's world model beliefs. Let $G = \{g_1, \ldots, g_n\}$ be the set of local goals of the agents in A. A multiagent planning problem from the perspective of a_i is a triple $M_i = (A, S_i, G)$.*

Representation of goals. In Section 3.5.2, it was shown that an agent's local goals are represented in a subgoal hierarchy. Hence, the representation of goals that is used to describe a MAPP is not flat, but rather describes a *path* within this goal hierarchy leading from the most specific subgoal that is relevant for the joint plan, up to more general subgoals, possibly to the top-level goal. Top-level goals, however, are the root entries of the plan stacks at the local planning layer. Thus, strictly speaking, the goal state of a MAPP is defined by the *intentions* of the participating agents rather than by their *goals*; each of the goals in Definition 3.6.2 describes a path in the plan hierarchy to the execution of which the corresponding agent has committed itself. This representation is partially due to the low expressiveness of the local planning mechanism described in Section 3.5.2, which does not rely on a state-based representation of goals: if a mechanism for planning from first principles were available, explicit representations of the goal states of the individual agents could be used to find a joint plan whose post condition actually implies the union of the goals. Therefore, although we shall stick to the notion of *goals* and *goal paths* for the sake of generality, the reader should keep in mind that the data structures that are used at the cooperative planning layer to test the applicability of joint plans are actually parts of the agents' intentions. Given the plan language $\mathcal{L}_0$ (see Definition 3.5.1) and the plan expansion function *expand* defined on page 84, a goal path is formally described as follows:

Definition 3.6.3. *Let $a_i \in \mathcal{A}$ be an agent. Let PL_i be the local plan library for agent a_i. A goal path g_i for an agent $a_i \in \mathcal{A}$ is either*

- *the empty plan $\epsilon \in \mathcal{L}_0$, or*
- *a conjunctive $\mathcal{L}_0$-plan $(g_{i1}, \ldots, g_{ik})$ with $g_{i(j+1)} \in expand(g_{ij}, PL)$ for each $1 \leq j \leq k - 1$.*

$(g_{i1}, \ldots, g_{ik})$ describes a path in the agent's plan stack, such that g_{ik} is a subgoal of $g_{i(k-1)}$, which is a subgoal of $g_{i(k-2)}$, aso.

The semantics of a goal $g_j = (g_{j1}, \ldots, g_{jk})$ in the definition of a MAPP is given as follows: firstly, agent a_j wants to achieve the goals given by the goal path $g_{j1} \ldots g_{jk}$; secondly, it cannot achieve goal g_{jk} without coordination with other agents.

We shall call the leaf node goal g_{jk} of the agent's goal path a *blocked goal*. It is of particular importance from an operational point of view, as it triggers the recognition of the need for cooperative planning. Blocked goals are, e.g., the goal to move to a place that is occupied by another agent, or the goal to lift a table, which is not possible without the assistance of another agent.

Next, we shall specify under what conditions a joint plan is applicable to solve a MAPP. For this purpose, we need to define the notion of dominance among two goal paths. Intuitively, one goal path g_1 dominates another goal path g_2 if g_1 ends with an instantiation of g_2:

Definition 3.6.4 (Dominance of goal paths). *Let g_1 and g_2 be goal paths, $g_1 = (g_{11}, \ldots, g_{1k})$, $g_2 = (g_{21}, \ldots, g_{2l})$. Then, g_1 dominates g_2 with substitution θ, written dominates(g_1, g_2, θ), if*

- *$l \leq k$ and*
- *for each $0 \leq j \leq l - 1$, $g_{1(k-j)} = g_{2(l-j)}\theta$.*

For example, the following relationships hold:

$$dominates((g_1(a), g_2, g_3(b)), (g_2, g_3(Y)), \{Y \leftarrow b\}),$$
$$\neg dominates((g_2, g_3(c)), (g_1(b), g_2, g_3(X)), \_).$$

In the following paragraph, the notion of dominance is used to define the applicability of joint plans to MAPPs.

Joint plans and joint plan libraries. Joint plans solve MAPPs: given a MAPP $M_i = (\mathcal{A}, S_i, G)$, a joint plan describes a set of single-agent plans whose coordinated execution leads to a world state S' in which the goals of the agents in $\mathcal{A}$ are no longer blocked: we say that the execution of the joint plan *unblocks* the agents' blocked goals specified in the MAPP. Note that we do not demand that a joint plan *achieves* the blocked goals, but only state the weaker requirement that the execution of the joint plan results in a world state that allows the individual agents to continue to achieve their local goals.

However, in some cases, joint plans can also achieve goals, which is reflected in the definition below.

The planning approach we have chosen is one from second principles, representing cooperative domain plans as entries in a library of joint plans, very similar to the treatment of single-agent plans in Section 3.5.2.

Definition 3.6.5 (Joint plan library). *A joint plan library is a collection of joint plans, i.e., a list of entries $JPL = [e_1, \ldots, e_k]$. The entries $e_i, 1 \leq i \leq k$ are of the form $e_i = (sit_i, res_i, ach_i, body_i)$, where*

*   *sit_i is the applicability condition of e_i;*
*   *res_i is the resolves–condition of e_i, i.e., a formula*

$$res_i \stackrel{\text{def}}{=} \bigwedge_{a_j \in A} blocked(a_j, g_j),$$

where g_j is a goal; we write $res_i(j)$ for the part g_j of res_i;
*   *ach_i is the achieves–condition of e_i, i.e., a formula*

$$ach_i \stackrel{\text{def}}{=} \bigwedge_{a_j \in A} achieves(a_j, h_j)),$$

where h_j is a goal; we write $ach_i(j)$ for the part h_j of ach_i;
*   *$body_i$ is the plan body of e_i; we use $body(e_i)$ to access $body_i$.*

The three conditions *sit*, *res*, and *ach* in Definition 3.6.5 are used to define under what circumstances a joint plan is applicable for solving a MAPP.

Definition 3.6.6 (Applicability). *Let JPL be a joint plan library. Let M be a MAPP, $M = (A, S, (g_1, \ldots, g_n))$. A joint plan $P = (sit, res, ach, body)$ is applicable to M with a substitution θ, written applicable(P, M, θ), if*

1. *there is a substitution θ_1 such that $sit\theta_1 = S$ and*
2. *for all $1 \leq j \leq n$, for $g_j = (g_{j1}, \ldots, g_{jk_j})$, and a substitution ρ_j, there is $g' \in res(j)$ such that $g_{jk_j} = g'(\theta_1 \circ \rho_j)$; let $\theta_2 \stackrel{\text{def}}{=} \theta_1 \circ \rho_n \circ \ldots \circ \rho_1$;*
3. *for all $1 \leq j \leq n$, and for $ach(j) = h_j$, there is a substitution τ_j such that dominates$(g_j, h_j\theta_2, \tau_j)$. Then, $\theta \stackrel{\text{def}}{=} \theta_2 \circ \tau_n \circ \ldots \circ \tau_1$.*

The intuition behind Definition 3.6.6 is the following: a joint plan P is applicable to a MAPP if (i) its applicability condition matches the situation description of the MAPP, if (ii) the leaf node goal of each agent matches a goal in the *resolves*–condition, and if (iii) for each agent a_j, the goal description g_j in the MAPP dominates the goal description p_j stated as a goal path in the *achieves*–condition of P. The second condition in Definition 3.6.6 ensures that a plan is applicable only if the goals specified in the MAPP are blocked; the third condition guarantees that—for each agent involved—an applicable plan does not achieve any goal that the agent currently does not have. We refer to Section 4.3.5 for examples.

Now, given a MAPP $M = (\mathcal{A}, S, G)$ and a joint plan library JPL, the access to the joint plan library is defined by a function *select*, which chooses those plans that are applicable to solve a given MAPP:

$$select(M, JPL) \stackrel{\text{def}}{=} \{P\theta | P \in JPL \wedge applicable(P, M, \theta)\}.$$

While the structure of joint plan libraries is very similar to that of single-agent plan libraries, there is an important difference in how they are processed. Whereas the latter (see page 84) are interpreted by means of a function *expand*, which defines hierarchical plans in a sense that abstract plan steps are refined recursively, the former are flat structures, which are transformed into single-agent plans (see below) to be expanded at the local planning layer.

Representation of joint plans. In the following, we define two languages describing the bodies of joint plans. Both languages are based on the language $\mathcal{L}_0$ for single-agent plans defined in Section 3.5.2 (see Definition 3.5.1), in that the parts of the agents in the joint plan are modeled as $\mathcal{L}_0$–plans. They are temporally related by means of *precedence constraints*.

Definition 3.6.7 (Plan language $\mathcal{L}_1$). *Let $\mathcal{A} = \{a_1, \ldots, a_n\}$ be a set of agents. An $\mathcal{L}_1$–joint plan JP is a structure $JP \stackrel{\text{def}}{=} [P, C]$, where P is either*

- *the empty plan ϵ, or*
- *a set $[(a_1, P_1), \ldots, (a_n, P_n)]$ of conjunctive $\mathcal{L}_0$–plans $P_i = p_{i1}, \ldots, p_{im_i}$, where each element (a_i, P_i), $a_i \in \mathcal{A}$, denotes that plan P_i is executed by agent a_i.*

$C = [c_1, \ldots, c_m]$ is a set of precedence constraints of the form $p_i < p_j$ between $p_i = p_{ik_i} \in P_i$ and $p_j = p_{jk_j} \in P_j$, $i \neq j$.

An $\mathcal{L}_1$-joint plan describes a set of single-agent plans, one for each agent involved, plus a set of precedence constraints restricting the possible courses of execution of the joint plan.

The plan language $\mathcal{L}_1$ directly assigns plans to agents; often, however, it is useful to introduce the concept of *roles* as an abstraction. In Section 3.6.2, roles have been defined in negotiation protocols. Analogously, a role in a joint plan describes a generic behavior within that plan. Roles in joint plans are instantiated during the process of joint plan negotiation described in Section 3.6.2 (see also Section 3.6.4 for the operational aspects).

The joint plan language $\mathcal{L}_2$, which is described in Definition 3.6.8, differs from $\mathcal{L}_1$ in that the individual single-agent plans are not labeled by agents, but by roles.

Definition 3.6.8 (Plan language $\mathcal{L}_2$). *Let $\mathcal{A} = \{a_1, \ldots, a_n\}$ be a set of agents. Let $\mathcal{R} = \{R_1, \ldots, R_k\}$ be a set of roles, let $r_1, \ldots, r_k$ denote role variables. An $\mathcal{L}_2$–joint plan JP is a structure $JP \stackrel{\text{def}}{=} [P, C]$, where P is either*

– *the empty plan ϵ, or*
– *a set $[(r_1 : R_1, P_1), \ldots, (r_k : R_k, P_k)]$ of conjunctive $\mathcal{L}_0$–plans $P_i = p_{i1}, \ldots, p_{im_i}$, where each element (a_i, P_i), $a_i \in \mathcal{A}$, denotes that plan P_i is executed by agent a_i.*

$C = [c_1, \ldots, c_m]$ *is a set of precedence constraints of the form $p_i < p_j$ between p_i in P_i and p_j in P_j, $i \neq j$. The precedence relation $<$ is a partial ordering (i.e., irreflexive, antisymmetrical, and transitive). It is an extension of the linear ordering induced by the single-agent plans. The plans P_i are allowed to contain occurrences of unbound role variables.*

However, when selecting a plan from the plan library, all role variables of that plan must be instantiated with names of agents by means of a role assignment function $\rho : \mathcal{A} \mapsto \mathcal{R}$. In general, ρ (see also page 101) describes an $n : m$ relation, i.e., one agent may play different roles in a plan, and one role can be played by different agents. In the examples discussed in this book, however, there are only $n : 1$ relationships, i.e., roles can be played by different agents (e.g., the role of a bidder in the contract net protocol), but we do not have the case that one agent plays more than one role in the examples investigated so far.

By using the concept of roles, plan language $\mathcal{L}_2$ allows to define joint plans in a more convenient and abstract manner than $\mathcal{L}_1$. However, the representation of precedence constraints for $\mathcal{L}_2$–plans is restricted to those between different roles; $\mathcal{L}_2$ does not allow us to formulate precedence between different agents playing the same role, as there is only one plan to represent each role.

Transformation into single-agent plans. Joint plans are meta-structures, which are not directly executable. Given the set $\mathcal{JP}$ of joint plans, and the set $\mathcal{P}$ of single-agent plans, the operational semantics of a joint plan is defined by a transformation function $transform : \mathcal{JP} \mapsto \mathcal{P}^n$, which maps a joint plan into n single-agent plans. The problem of joint plan transformation can be formulated as a graph-theoretic problem. For a joint plan $JP = [P, C]$, a graph Gr_{JP} is generated as follows: first, for each single-agent plan $P_i = [p_{i1}, \ldots, p_{im_i}]$, a directed linear subtree is generated, where the root node is p_{i1} and where p_{ij_i} is the successor of $p_{ij_{i-1}}$. Then, for each pair of nodes p_{il_i}, p_{jl_j} in different subtrees for which there is a precedence constraint $p_{il_i} < p_{jl_j}$, a directed edge from p_{il_i} to p_{jl_j} is introduced. Figure (3.26.a) shows the graph Gr_{JP} corresponding to a joint plan

$$JP = [[[a_1, a_2, a_3], [b_1, b_2], [c_1, c_2]], [a_1 < b_2, b_2 < c_1, c_2 < a_3]].$$

The transformation function $transform$ consists of three *transformation rules*, TR1, TR2, and TR3:

TR1: For each node n_i in a subtree s with an edge e emanating to a node m_j in a different subtree t, the subtree s is modified by inserting a new node $\bar{n}$ labeled $send(t)$ directly behind n_i, and by marking edge e.

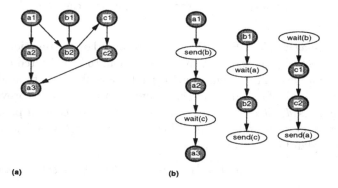

Fig. 3.26. Graph representation of a joint plan (a) before and (b) after transformation

TR2: For each node m_j in a subtree t with an incoming edge e from a node n_i in another subtree s, subtree t is modified by inserting a new node $\bar{m}$ labeled $wait(s)$ directly before m_j, and by marking edge e.

TR3: Delete any edge e that is marked twice.

The effects of the modification rules are graphically illustrated in Figure 3.27. The graph representation of the result of applying the function

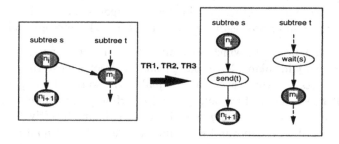

Fig. 3.27. Joint plan transformation rules

transform to the above plan JP is shown in Figure (3.26.b). The plan steps send(<agent-name>) and wait(<agent-name>) are synchronization commands that are implemented as procedure PoBs in the behavior-based layer:

- **send(a)** sends an acknowledgment message, i.e., a message with type IN-FORM and content ok to the agent specified by a.
- **wait(a)** waits until an acknowledgment message is received from agent a. Wait is synchronous; however, a time-out mechanism is built in to prevent deadlocks. If no such message is received within the time interval specified

for the time-out, `wait` returns `fail`, which causes the joint plan execution to fail.

Given single-agent plans $P = (P_1, \ldots, P_n)$ returned by the function *transform*, we use $project_i(P)$ to access the single-agent plan to be executed by agent a_i.

The above graph representation for joint plans allows us to analyze them by using the tools of graph theory. For example, it is possible to prove whether a joint plan is *deadlock-free*, i.e., whether it does not contain plan steps p_1 and p_2 with $p_1 < p_2$ and $p_2 < p_1$. Theorem 3.6.2 states a sufficient criterion for a joint plan being deadlock-free:

Theorem 3.6.2. *A joint plan JP is deadlock-free iff the corresponding graph Gr_{JP} is acyclic.*

Evaluation of joint plans. The decision which suitable joint plan to prefer in a negotiation is based on a model allowing an agent to evaluate joint plans by computing their utility. The model proposed in the following is based on the definition in Section 3.5.3 for single-agent plans: the utility of plans is defined as the difference between their worth and their costs. However, computing the utility for joint plans is more difficult:

1. Joint plans are not directly executable structures; therefore they must be transformed into single-agent plans before they can be evaluated, taking the cost of synchronization into account.

2. As multiple agents are involved in joint plan negotiation, different utility functions cause different group behavior. If each agent only takes into account the utility of its part in the plan, agents are individual rational, and the result of the negotiation will be at most Pareto-optimal [vNM44].

An additional problem, which we omitted in the single-agent case, is the following: two joint plans, let us say p_1 and p_2, may achieve the same goal g of an agent. As the worth of a plan is equal to the worth of the goal achieved by the plan, we should have $w(p_1) = w(p_2)$. However, assume that plan p_2 additionally achieves the goal g' of which g is a subgoal. Intuitively, the worth function should reflect this and we should have $w(p_2) > w(p_1)$ if p_2 satisfies a more general goal than p_1. In the following, we shall define a utility function for joint plans that satisfies these requirements. First, we recall the basic definition of a utility function:

Definition 3.6.9 (General utility function for joint plans). *Let $\mathcal{A}$ be the set of agents. Let $\mathcal{JP}$ be a set of joint plans. Then $u : \mathcal{A} \times \mathcal{JP} \mapsto \mathbb{R}$ is the utility function for joint plans. The utility of a plan $p \in \mathcal{JP}$ for agent $a_i \in \mathcal{A}$ is computed as*

$$u(a_i, p) = w(a_i, p) - c(a_i, p),$$

where $w : \mathcal{A} \times \mathcal{JP} \mapsto \mathbb{R}$ is a worth function, and $c : \mathcal{A} \times \mathcal{JP} \mapsto \mathbb{R}$ is a cost function for joint plans.

Definition 3.6.9 easily extends to plan languages that use the concept of roles: in this case, we have $u : \mathcal{R} \times \mathcal{JP} \mapsto \mathbb{R}$, and $u(r, P)$ specifies the role-specific utility of a joint plan $P \in \mathcal{JP}$ relative to role $r \in \mathcal{R}$.

Costs of joint plans:. These are determined by evaluating the transformed plans using the single-agent cost function c_0 (see Definition 3.5.3). We distinguish between two different types of cost functions reflecting two agent types: local cost functions for *self-interested agents* and global cost functions for *cooperative agents*.

Definition 3.6.10 (Local cost function). *Let $P \in \mathcal{JP}$ be a joint plan for a set $\mathcal{A}$ of agents, $a \in \mathcal{A}$. Let $\bar{c}_0$ be the cost function c_0 from Definition 3.5.3 extended by cost assignments for the synchronization actions* wait *and* send. *Then,*

$$c_1(a, P) = \bar{c}_0(\text{project}(a, \text{transform}(body(P))))$$

is a local cost function denoting the cost of plan P for agent a.

Function c_1 computes the cost of a joint plan for an agent as the cost of the single-agent plan the agent performs within the joint plan. The definition uses the functions *project* and *transform* defined above (see page 112). Thus, when evaluating a joint plan using cost function c_1, an agent only takes into account its role in the plan.

If the application under consideration permits both the assumption of cooperative agents that act according to a global utility function, and the availability of a global cost function, the cost function for joint plans can be defined in a different way:

Definition 3.6.11 (Global cost function). *Let $\mathcal{JP}$, $\mathcal{A}$, $\bar{c}_0$ as in Definition 3.6.10. Then, for each $a \in \mathcal{A}$,*

$$c_2(a, P) = c_2(P) = \sum_{a_j \in \mathcal{A}} \bar{c}_0(\text{project}(a_j, \text{transform}(body(P)))).$$

is a global cost function for joint plans.

This function computes the cost of a joint plan as the sum of the costs of all its single-agent projections. There are some prerequisites to such a function: the joint plan must be known to all agents, and agents need to know each other's local utility functions. In Definition 3.6.11, we assume that the cost function for single-agent plans is identical for each agent. This assumption makes computing c_2 simple; it is reasonable if we deal with homogeneous agents. It can be weakened by indexing the single-agent cost function by the name of the agent in the heterogeneous case.

Worth of joint plans. The worth of a joint plan is defined as the worth of the outcome of the plan. This outcome is given by the *resolves*–condition of the plan (i.e., the goals that are unblocked by the plan) plus the *achieves*–part of the plan (i.e., the goals that are achieved by the plan). As the goals in these conditions are formulated using goal paths (see Section 3.6.3), we define the worth of a joint plan with respect to goal paths. As in the case of cost functions, we make a difference between the local and the global worth of a joint plan. We start by defining a worth function over goal paths.

Definition 3.6.12 (Worth function for goal paths). *Let $g = (g_1, \ldots, g_k)$ be a goal path. Let $\hat{w}$ be a function that maps each g_j, $1 \leq j \leq k$ into a real number κ: $\hat{w}(g_j) = \kappa$, with $1 \leq j < l \leq k \rightarrow \hat{w}(g_j) > \hat{w}(g_l)$. Then, $w_0(g) \stackrel{\text{def}}{=} \hat{w}(g_1)$ denotes the worth of the goal path g.*

Thus, the worth of a goal path is given by the worth of its root node, i.e., of the top-level goal of that path. The requirement $1 \leq j < l \leq k \rightarrow \hat{w}(g_j) > \hat{w}(g_l)$ reflects the intuitive criterion that the worth of a goal should be bigger than the worth of any of its subgoals. Whether we can make stronger assumptions like superadditivity ($\hat{w}(g_1 \wedge g_2) \geq \hat{w}(g_1) + \hat{w}(g_2)$) depends on the domain under consideration.

Next, we define worth functions for joint plans:

Definition 3.6.13 (Local worth function). *Let $P = (sit, res, ach, body) \in \mathcal{JP}$ be a joint plan. Let $a_i \in A$ be an agent participating in P; let $res(i) = g$ and $ach(i) = h$ be the goal paths for a_i specified by the resolves- and achieves–conditions of P. Let ϵ_1, ϵ_2 be real numbers, $\epsilon_1, \epsilon_2 \in [0, 1]$. Then,*

$$w_1(a_i, P) = \epsilon_1(a_i) \cdot w_0(g) + \epsilon_2(a_i) \cdot w_0(h)$$

is a local worth function denoting the worth of plan P for agent a_i.

An agent that uses function w_1 for plan evaluation computes the worth of a joint plan as the weighted sum of the worths of those of its local goals that are unblocked and of those that are achieved by the plan. The weights ϵ_1 and ϵ_2 are functions of the respective agent. Thus, the agent only takes into account its part in the plan. Whereas this is reasonable in the case of self-interested or even adversarial agents, a worth function for joint plans in a cooperative domain can take into account the global worth of a plan. This requirement is reflected in Definition 3.6.14:

Definition 3.6.14 (Global worth function). *Let P, ϵ_1, ϵ_2, $res(i)$, $ach(i)$ be as in Definition 3.6.13. Then, for each $a_j \in A$,*

$$w_2(a_j, P) = w_2(P) = \sum_{a_i \in \mathcal{A}} \epsilon_1(a_i) \cdot w_0(res(i)) + \epsilon_2(a_i) \cdot w_0(ach(i))$$

is a global worth function for joint plans.

Based on the worth and cost functions defined above, we can refine the definition of a utility function as follows:

Definition 3.6.15 (Utility functions for joint plans). *Let A be a set of of agents. Let $\mathcal{JP}$ be a set of joint plans. Then, for $a \in A$ and $P \in \mathcal{JP}$,*

1. *$u_1(a, P) = w_1(a, P) - c_1(a, P)$ is a function denoting the local utility of plan P for agent a.*
2. *$u_2(a, P) = u_2(P) = w_2(P) - c_2(P)$ denotes the global utility of plan P for agent a.*

The global utility of a joint plan is equal for all agents participating in the plan.

Theorem 3.6.3. *Let $A = \{a_1, \ldots, a_n\}$ be a set of agents, let $\mathcal{JP}$ be the set of joint plans with agents in A. Then, for each $P \in \mathcal{JP}$ and for each a_i, $a_j \in A$, we have $u_2(a_i, P) = u_2(a_j, P)$.*

In Section 4.3.5, we give examples for utility functions used in the loading dock.

3.6.4 The control cycle

The control cycle is triggered by activation requests from the local planning layer either if a conflict occurs that cannot be solved locally, or if a task is assigned to the agent for which cooperation might be appropriate. The control cycle of the cooperative planning layer is an instance of the generic layer control cycle shown in Figure 3.8. It is implemented by the method `cycle` shown in Figure 3.28. The top-level control loop of the cooperative planning layer is almost identical to that of the local planning layer (see Figure 3.17): whereas the latter manipulates plan stacks, the former maintains protocols. In each loop, first, the messages received from the local planning layer are processed (see Figure 3.29). This causes the creation of new protocol stacks and the modification of existing stacks. Thus, this function implements situation recognition and goal activation.

In the second step of the control loop (lines 23–30), each existing protocol stack is processed. As it is possible for the local planning layer to maintain multiple goals by handling multiple plan stacks, the cooperative planning layer can carry out multiple negotiation processes. Each stack is processed by interpreting its top element. The interpretation of a protocol is described below (see Figure 3.30). It returns the next step of the protocol, which has to be integrated into the schedule (line 27); then, the execution function is called scanning the schedule for actions that can be executed. In the following, the steps of the control cycle are discussed in detail.

```
/* Definition of the class CPL. The methods process-msg, interpret */
/* are described in detail in Figures 3.29 and 3.30, respectively */

1 class CPL
2 super layer
3    attributes
4       Prot-stacks      /* list of current protocol stacks */
5       Msgs      /* inter-layer message queue */
6       Prot-sched      /* current schedule of protocol execution*/
7       Jp-lib     /* joint plan library */
8       Prot-lib      /* negotiation protocol library */
9       Strat-lib      /* library of negotiation strategies*/
10      [...]
11   methods
12   [...]
13      meth process-msg <?Msg>
14      meth generate-jplan <?Goal-descr>
15      meth evaluate-jplan <?Jplan>
16      meth interpret <?Prot-stack>
17      meth schedule <?Step>
18      meth execute
19   [...]
20      meth cycle
21         foreach M ∈ Msgs do
22            process-msg(M); od
23         foreach Ps ∈ Prot-stacks do
24            if pending(Ps) then true
25            else
26               Step = interpret(Ps);
27               Sched = schedule(Step);
28            fi
29            Sched = execute(Sched);
30         od
```

Fig. 3.28. The control cycle of the cooperative planning layer

Situation recognition and goal activation. These functionalities are implemented by the function `process-msg` which processes the messages received by the local planning layer. Figure 3.29 illustrates this function, which is a method of the CPL object.

The most complex case occurs if an upward activation request do((S,G), Args) is achieved (lines 3–12). In this case, the cooperative problem-solving process is initiated according to the general negotiation model defined in Section 3.6.2. According to this model (see also Figure 3.21), the layer first tries to obtain additional information about the problem (lines 4–7) by retrieving information about all other agents' goals. Based on this information, the situation is classified. This corresponds to the definition of the MAPP, consisting of the external context (the argument Sit in Figure 3.29), the mental context

```
/* method process-msg of class CPL; ps(Id) is the access function */
/* to the protocol stack created by the request with id = Id. Sdr, */
/* id are access functions to the sender and the Id of a message; */

1   meth process-msg(M)
2     case Msg of
3       'do((Sit, G), Args)':
4           goals= own-goals(Sit,G)
5           foreach Agent ∈ Participants do
6              Agent←query(curr-goal(Sit)) od
7           O-goals=collect-replies(Participants);
8           Ctype=classify(Sit,Goals,O-goals);
9           if Ctype ≠ nil then
10              create-prot-stack(negotiate(Ctype,Sit,Goals,O-goals);
11          else
12              lpl←inform(done(id(Msg), no-conflict)); fi

13      'stop(Id)':
14          rm-from-schedule(ps(Id), Sched);
15          rm-prot-stack(ps(Id), Prot-stacks);

16      'done(Id, Stat)':
17          Status = determine-status(Id)
18          if Status = succ then
19              cleanup(ps(Id));
20          else replan(ps(Id)); fi

21      'eval(Jp)':
22          U = evaluate-plan(Jp);
23          lpl←inform(done(id(Msg), U))

24      'interpret(Jp)':
25          create-prot-stack(transform(Jp));
26    esac
```

Fig. 3.29. Processing inter-layer messages in the CPL cycle

Goal, and the social context O-goals. If there is a conflict that should be dealt with at the cooperative planning layer, a new protocol stack is created and initialized with the goal negotiate(Ctype,Sit,Goals,O-goals), which is processed by the protocol interpretation function (see Figure 3.30).

Another message that can be received from the local planning layer is the request to stop a currently active negotiation, e.g., because the local situation of the agent has changed in a way that has made the current coordination process obsolete. In Figure 3.29, this case is treated by removing both all actions that are already scheduled and the protocol stack. However, stopping a coordination process based on local considerations is a critical issue in general, as during the coordination process, commitments are made to other agents that should not be broken. E.g., a robot should not decide to pick up a new box while holding a box together with another robot. Even if the breaking of a commitment is uncritical, clean-up protocols are needed to finish the cooperation process in a decent way.

Further activities of the cooperative planning layer that are treated by the function `process-msg` are the processing of an acknowledgment from the local planning layer (lines 16–20), the evaluation of an existing joint plan (lines 21–23, see also the discussion below), and the interpretation of a predefined joint plan (lines 24–25).

Protocol interpretation. Each time the control cycle is executed, the function `interpret` (see Figure 3.30) processes the protocol step on top of each protocol stack. Multiple stacks have been chosen as the interpretation and execution model for negotiation protocols as they provide a simple model for stepwise and concurrent execution of multiple protocols.

Different possible cases are handled in the interpretation function. Lines 6–16 implement the phase model of negotiation defined in Section 3.6.2, indicated by the expression `negotiate(Ctype,Sit,Goal,O-goals)` on top of the stack: first, a protocol is selected, and the roles are assigned to the participants using an election protocol (see page 99). Depending on its role, each agent selects a strategy from the strategy library. The winner of the election creates the negotiation set. The actual plan generation method `generate-jplan` (see Figure 3.28) is called from the function `compute-neg-set`. The negotiation set is broadcasted to all participants, and the actual negotiation starts by pushing `exec-prot((Prot,Role,Str,Ns), start))` on the protocol stack. Different actions within this procedure, such as the role assignment and the broadcasting of the negotiation set, require the execution of protocols. They are initiated by pushing corresponding actions on the stack (lines 33–34). The execution of a protocol is initiated by the stack content

```
exec-prot((Prot,Role,Str,Ns), State),
```

where `Prot` is the protocol name, `Role` is the role the agent plays in the protocol, `Str` is the strategy that the agent has selected, `Ns` is the negotiation set, and `State` is the current state of the protocol (= `start` at the beginning). It is shown in lines 17–29. First, the alternatives in the protocol are computed. Then, the next action is selected from among these alternatives by using the current strategy. If the protocol execution is finished successfully returning a joint plan, this plan is transformed into a single-agent plan and the corresponding commitment is sent to the local planning layer (lines 30–32). Otherwise, the selected action is returned as a result of the function call, the negotiation set is updated according to the negotiation strategy (see Section 3.6.2), and the negotiation continues.

Protocol scheduling and execution. In each cycle, the output of interpreting the individual protocol stacks is collected and scheduled by the protocol scheduler. The protocol scheduler also classifies outgoing and incoming messages according to the negotiation context to which they belong and thus

```
/* Input argument Ps denotes a plan stack; the stack operators */
/* empty, push, pop are defined as usual */

1  meth interpret(Ps)
2   if empty(Ps) then return true; /* no protocols to execute */
3   else
4     Top = pop(Ps); /* get top of stack */
5     case Top of
6       'negotiate(Ctype,Sit,Goal,O-goals)':
7         select-protocol(Ctype,Sit,Goal,O-goals,Prot-lib);
8         push(Ps, [
9           assign-roles(Goal,O-goals,Elect-fnc,Prot,Role-assg);
10          select-strategy(Goal,O-goals,Prot,Role-assg,Strat-lib,Str);
11          if Role-assg=leader then
12            compute-neg-set(Ctype,Sit,Goal,O-goals,Jp-lib,Ns);
13            broadcast(Ns, Participants);
14          fi;
15          push(Ps, exec-prot((Prot,Role,Str,Ns), start));
16          ]);
17      'exec-prot((Prot,Role,Str,Ns), State)':
18        Prot-alts=π(Prot,Role,State)
19        Next-act=σ(Role,Prot-alts);
20        if next-act=(done,Jp) then
21          push(Ps,transform(Jp)); return true; fi
22        else
23        if Next-act=(fail) then return failure fi
24        else
25          Ns=compute-negset(Next-act,Ns);
26          Newstate=update-neg-state(Prot,Role,State,Ns,Next-act);
27          push(Ps,'exec-prot((Prot,Role,Str,Ns),Newstate)');
28          return Next-action;
29        fi
30      'transform(Jp)':
31          P=project(transform(Jp));
32          return lpl←commit(P);
33      'assign-roles(...)', 'select-strategy(...)',
34      'compute-neg-set(...)',...:/* execute protocols of Sec. 3.6.2 */
35    esac
36 fi
```

Fig. 3.30. The protocol interpretation function

prevents confusion between messages received in different negotiations. It also splits broadcast messages into sets of messages, one to be sent to each agent specified in the parameter list of the broadcast command. In the current version of the cooperative planning layer, the scheduler is simple, as there are no real constraints between different negotiations that are carried out concurrently. An additional functionality of the scheduler, which is currently not part of the system, can be to detect cycles in higher-level protocols for task allocation, where e.g., an agent is announced an order that itself announced before.

The actual execution mechanism makes sure that actions scheduled for execution are executed in time by posting the corresponding commitment message down to the local planning layer. Additionally, it monitors time-outs that are specified for the negotiation protocols and triggers an event if no reply has been received after the time specified in the time-out.

3.6.5 Interfaces

There are interfaces between the cooperative planning layer and the agent knowledge base as well as the local planning layer. The relationship between the cooperative and the local planning layer was discussed in Section 3.5.4; we refer to that section for details.

The cooperative planning layer has access to all layers of the agent knowledge base: world model information is accessed to obtain information about the current state of the world. The layer can retrieve information regarding the agent's current goals via the mental model. As the activation of the cooperative planning layer is bottom-up, i.e., from lower layers, the layer receives basic information about a current situation and the agent's goals that are affected by this situation from the local planning layer (which itself obtains information extracted from the agent's world model from the behavior-based layer). Thus, the cooperative planning layer obtains abstracted knowledge characterizing, e.g., conflict situations; situations are pre-classified by their external and mental context. At the cooperative planning layer, this description of an interaction situation is enhanced by information about the goals of other agents that are involved in the situation. This information is provided from the social model (see Section 3.3.3). On the basis of this information, the cooperative planning layer classifies interaction situations, selects protocols, and maintains their execution.

In summary, the cooperative planning layer of an agent provides facilities allowing an agent to coordinate its actions with other agents by negotiating and executing *joint plans*. Its activity is triggered by requests from the local planning layer; its output are single-agent plans that describe the agent's role in a joint plan. We have described a negotiation model for INTERRAP agents, and we have defined an approach to multiagent planning from second principles using a library of domain plans.

3.7 Bottom Line

An INTERRAP agent consists of a knowledge base, a world interface, and three hierarchically organized control layers: (i) the behavior-based layer provides basic reactive behavior and procedural knowledge for routine tasks; (ii) the local planning layer incorporates the agents capabilities to do means–ends reasoning to achieve its local goals; (iii) the cooperative planning layer holds facilities allowing an agent to coordinate its actions with other agents through the negotiation of joint plans.

4. The Loading Dock: A Case Study

In this chapter, we give an example of an agent-based system designed on the basis of the previously described architecture. We show how the architecture supports the design of agents that are at the same time reactive, deliberative, *and* capable of interaction.

4.1 Introduction

The following application models an automated loading dock. Figure 4.1 illustrates the domain: The loading dock is a rectangular area, which contains shelves and a loading ramp. There are different types of boxes, marked by different colors; boxes with a certain color must be stored on shelves of the corresponding color. The loading dock is inhabited by forklift robots. These

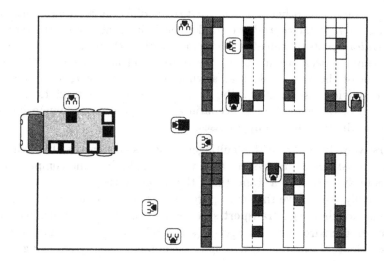

Fig. 4.1. The loading dock domain

robots carry out certain tasks: e.g., unloading boxes from the ramp and storing them on the shelf or vice versa. Each forklift is able to carry exactly one box at a time.

Scenarios like this are a reality in many manufacturing sites. For example, at the Cologne plant of the car manufacturer FORD, more than 40 flexible transportation systems implement the flow of material through the areas of production and car assembly. These systems measure about 2 meters in length and 1.50 meters in height each; they follow magnetic induction lines installed in the floor and receive their orders from a central computer. For safety reasons, they are equipped with ultrasonic sensors and bumpers. Collisions between different systems are prevented by a semaphore–like strategy: the driving zone is divided into several areas, each of which may only be used by one system at a time. The navigation task is solved by the central computer that guides the systems from landmark to landmark.

Today's flexible transportation systems have local sensory systems and a very limited amount of autonomy (e.g., drive from one landmark to another, load or unload a pallet). However, large amounts of data are still exchanged between the transportation systems, the machines that are served by them, and the central computer. Thus, the central computer is likely to become a bottleneck: failure of the communication network or of the central computer itself causes the whole system to crash involving high cost. Moreover, the fact that conflicts between systems such as blockings are avoided *a priori* at the design time of the system reduces flexibility: changes in the structure of the system require a global redesign, as the systems themselves do not scale up and are not capable of dealing with new situations due to their low degree of autonomy.

For these and other reasons, we take a different approach: instead of relying on a *global control*, we model the individual transport systems as *distributed agents* that deal autonomously with both local task planning and with conflicts that occur due to the presence of other agents. The overall behavior of the loading dock as a whole is then an *emerging functionality* of the individual skills of and the interaction among the forklifts. Thus, the selection of the loading dock scenario as the application domain has been made on the basis of two hypotheses:

- Firstly, we believe that the multiagent systems approach is able to increase the *robustness* of the system as a whole. Faults in some components are less likely to lead to a standstill of the whole system.
- Secondly, we believe that our approach increases the *flexibility* of the system. Modeling e.g., transport systems and machines in a flexible manufacturing environment as agents (see [Fis93b]) makes it much easier to reconfigure the system by adding new components or by removing existing ones.

Since the beginning of the 90s, the field of interacting robots has received much attention in robotics and multiagent systems research (see e.g., [Kan93]). However, most approaches are purely reactive and their scope is restricted to the modeling of simple grouping and tracking behaviors (see e.g., [Mat93] [Ste94] [BA95]. In the following, we shall look at a higher, symbolic level of research on interacting robots: the forklift agents in the loading dock are not only capable of moving around the loading dock, of perceiving their environment, and of taking boxes off the shelves or the ramp, and of storing them back there. They also construct internal symbolic representations of the world, devise plans to achieve their local goals, communicate with other robots, and have a whole spectrum of behavior-based and deliberative means at their disposal, which allow them to interact with other agents. While much research has been done in modeling (reactive and deliberative) robots in single-agent environments, to our knowledge, the loading dock application has been the first interacting robots application combining reactive mechanisms of interaction with those using cooperative planning.

The loading dock domain is a suitable testing environment for our agent design concepts for several reasons:

- The fact that several physical systems equipped with sensors and actors exist in a shared environment causes a high degree of dynamics: the world changes unexpectedly and the agents have to react to these changes in time.
- Agents are required to perform their local transportation tasks efficiently and should take their local goals into account in interactions with other agents.
- The lack of a global control requires a variety of mechanisms to deal with different types of interactions among the agents.
- The domain is sufficiently complex and dynamic so that it is not reasonable to model the decision behavior of the agents by using *optimal* methods, e.g., by using a theorem prover. Rather, agents are resource-bounded, and the control mechanism should account for that.
- On the other hand, the domain is sufficiently simple to be easily understood. It allows us to focus on some important aspects and to abstract away other, less important ones (see below).

Our focus is on the aspects of agent modeling, i.e., on the interplay of reactivity and deliberation and on mechanisms of interaction. This chapter focusses on the latter aspect, i.e., on how different types and methods of coordination among forklift agents are supported by the INTERRAP architecture; the relationship between reactivity and deliberation and the problems related to achieving coherent agent behavior in systems providing hybrid representations of these paradigms is the subject of future research.

The loading dock domain is sufficiently complex to enforce several restrictions and general assumptions to be made throughout this book:

- The structure of the environment is restricted. The ground is represented as a rectangular grid, and each object in the domain (shelves, ramp, boxes, forklifts) is rectangular and can be mapped into individual squares or to areas in the grid.

- As our focus is not on sensory data processing, we make the *symbolic sensor assumption* (see Section 3.2, page 50).
- We do not place emphasis on a sophisticated actoric system: While the detection of the absolute position of a robot is still an important research problem in robotics, we solve this problem by assuming that a robot is logically assigned to exactly one square in a grid at a time; once the robot knows its initial position, it can easily compute its current position relative to this initial position at any point in time. In the implementation of the loading dock using real robots, the robots orient themselves by means of white lines painted on the floor.
- We restrict ourselves to a specific type of interaction, namely on the resolution of conflicts between agents (also called *negative interaction* [Syc87]). In the transportation domain, we have investigated coordination between agents in the achievement of transportation tasks (i.e., *positive interaction*), where cooperation is regarded as a means of achieving local goals (see e.g., [FKM94] [FMP95b] [FMP96]).

The behavior of agents in the loading dock is modeled both in a software simulation *and* in an — still somewhat idealized — implementation on KHEPERA miniature robots (see also Section 4.4).

The structure of this chapter is as follows: in Section 4.2, the modeling of the individual forklifts as INTERRAP agents is described. The agent knowledge base, the world interface, and the individual control layers are instantiated. Section 4.3 defines the notion of conflicts in the loading dock and demonstrates several mechanisms of how conflicts can be recognized and solved by an INTERRAP agent. In Section 4.4, the simulation environment and the implementation with KHEPERA robots are outlined.

4.2 The Design of Forklift Agents

4.2.1 Knowledge representation

The beliefs of a forklift agent in the loading dock are modeled within the AKB formalism as described in Section 3.3.3. As the knowledge representation in AKB reads awkwardly, we introduce the following short-hand notation: an AKB concept definition of the form

```
concept(       name:       p
attribute(     name:       a₁
               concept:    p )
value(         name:       a₁
               val:        v₁ )
attribute(     name:       a₂
               concept:    p )
value(         name:       a₂
```

```
                        val:        v₂ )
      [...]
      attribute(        name:       aₙ
                        concept:    p )
      value(            name:       aₙ
                        val:        vₙ )
```

is written as

$$p(a_1:v_1 \quad a_2:v_2 \quad ... \quad a_n:v_n)$$

Type declarations are omitted in the examples given in this section. In the following, formulae of the form $p(a_1:v_1 \ a_2:v_2 \ ... \ a_n:v_n)$ are simply called *facts*.

The three levels of knowledge of an INTERRAP agent, i.e., its world model, its mental model, and its social model, are described in the following:

The world model. The world model of a forklift is represented by a collection of ground facts. It describes the object classes that form the agent's environment. Examples of object classes in the loading dock are square, region, agent, shelf, ramp, and box. E.g., each square of the loading dock is represented as a fact

```
square(oid:id xcoord:x ycoord:y type:t status:s).
```

For example, the fact

```
square(oid:$1423 xcoord:3 ycoord:5 type:ground status:nil
)
```

denotes that square (3,5) is a ground square which is currently free. The attribute oid denotes the object identification of the fact. References to objects in these facts are denoted by their object identifiers. For example, if we want to represent that forklift f_1 is on square (3,5) in the definition of the square, we can do so by writing square(oid:$1423 xcoord:3 ycoord:5 type:ground status:$4711), where $4711 is the object identifier of agent f_1. Thus, object identifiers can be used to implement foreign keys in a relational database scheme. This allows us to avoid nested facts, i.e., attributes whose values are themselves facts. We omit the oid attribute whenever it is not necessary in our examples.

Object-level knowledge about forklift agents is stored in facts of the form

```
agent(    type:... name:... address:... pos:(x y o)
          carries:... gripper:... hand:...),
```

where type denotes the agent type (there is only one agent type in the loading dock), name is the name of the agent, address is its communication address, pos is a triple containing its x- and y-coordinates, and orientation (s, w, n, or e), gripper denotes the existence/ position of the gripper, respectively (none, up, or down), and hands denotes the existence/status of the gripper hands (none, open, closed).

The beliefs stored in the agent's world model may be incomplete. This is denoted by missing values for attributes[1]. For example, a forklift agent might know about the existence of a specific green box b_{23} while not knowing the current location of b_{23}:

box(oid:$813 name:b$_{23}$ type:green xcoord:nil ycoord:nil).

Moreover, the beliefs of an agent may be incorrect with respect to the actual state of the world. For example, an agent may believe that box b_{23} is at location (7, 7), indicated by a fact box(name:b$_{23}$ type:green xcoord:7 ycoord:7), as it may have seen the box at this location at an earlier point in time; however, the current location of this box may be different as another agent may have displaced the box.

The mental model. This section of the agent knowledge base contains representations of the agent's local goals. As a mechanism for planning from second principles is used in the loading dock (see Section 3.5), goals are not represented as sets of facts, but as specific facts denoting the abstract operation which is required to satisfy the goal. This short-hand representation allows us to use goals as indexing elements for the plan library. An example for a goal fact is

goal(type:load-ramp name:r$_1$ box-id:$813),

denoting that the agent has the goal to load the box with the object id $813, which is the green box b_{23} in the above example. Basically, one goal fact is maintained for each top-level goal. The expansion of local goals into a goal hierarchy is maintained in the control part of the agent by means of different goal stacks. The principles of processing local goals in INTERRAP have been described in Section 3.5. Section 4.2.3 gives a detailed example.

The social model. This part of the knowledge base of a forklift holds the beliefs about other agents. Whereas object-level beliefs about others (e.g., a forklift f_1 may believe that another forklift f_2 is located at square (2,4)) logically belong to the world model part of the knowledge base, information that is relevant for cooperative planning (see Sections 3.6 and 4.3) is related to other agents' mental models. E.g., the beliefs of an agent f_1 about another agent f_2 are represented in f_1's world model by a fact

agent(type:forklift name:f$_2$ address:F2/serv-101/48372
         goals:$234 pos:(3,4,n) carries:$813 gripper:up hand:closed)

The object denoted by the object identifier $234 in the agent fact refers to the fact in the social model of f_1, in which its beliefs about f_2's goals

---

[1] In database systems, the semantics of missing values is an important topic, as in general, it is hard to tell whether a missing attribute value means that this value is undefined (i.e., does not exist), or whether it is not known. In our framework, we make the assumption that the attribute value nil means that the value is unknown.

are stored. We have seen in Section 3.5 that an agent can have different goals in parallel. On the one hand, different independent top-level goals can be pursued; on the other hand, an abstract top-level goal is expanded into subgoals. Thus, e.g., an agent which has the goal to load a specific box onto the loading ramp, can currently work on the specific subgoal of searching that box in a shelf, and on the sub-subgoal to move to a specific landmark to inspect a specific area of the shelf. Thus, the social model of an agent is hierarchically organized in the same way as the local goals of the agent itself.

This is accomplished as follows: for each other agent known to an agent, there is a fact o-goals, which contains references to all the top-level goals of that agent, e.g., o-goals(oid:$234 name:f$_2$ goals:[$75 $693 $715]). Each member of the goals list is the object identifier of a fact o-goal, e.g.,

o-goal(oid:$75 name:f$_2$ type:load-ramp name:r$_1$ box-id:$629
   subgoals:[$1001 $1093]).

Agent f_1 believes that agent f_2 has the top-level goal load-ramp. It also believes that f_2 has the subgoals denoted by the object identifiers $1001 and $1093 as parts of the subgoal. Thus, an agent can model other agents' goal hierarchies.

The social model of an agent may be both incomplete and incorrect. Incompleteness means that e.g., f_1's representation of agent f_2's mental model only represents a subtree of the actual goal hierarchy (see Figure 4.2) of a specific top-level goal, or that not all top-level goals of f_2 are known to f_1.

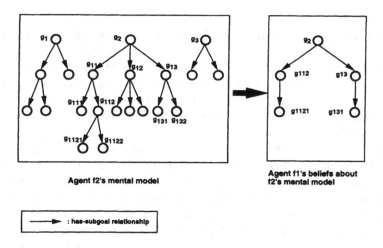

Agent f2's mental model

Agent f1's beliefs about f2's mental model

⟶ : has-subgoal relationship

Fig. 4.2. Incomplete beliefs in the social model

In the example, only goal g_2 of agent f_2 is represented in the social model of f_1. Moreover, f_1 does not know about the subgoal g_{12}. Additionally, subgoal information may be missing within one path. In the example, f_1 believes that

f_2 has the goal g_{112} as a subgoal of g_1, but f_1 does not know about goal g_{11} which is in between g_1 and g_{112}.

The social model may be incorrect as it may contain outdated information. As we assume throughout this work that agents will reveal their goals truthfully if asked (see Section 3.6), we can exclude incorrectness due to lying agents.

As we have seen in Section 3.6, multiagent planning problems are represented by a collection of goal facts, and entries of the joint plan library are indexed by such collections of goals. In Section 4.3, this shall be detailed by an example.

4.2.2 The world interface

The world interface of a forklift agent accounts for its capability of perceiving the environment, of receiving messages from and sending messages to other agents, and of carrying out actions. In the following, the general description of the world interface given in Section 3.3.2 is detailed by the example of the simulated forklift agents. Compared to the implementation using KHEP-ERA miniature robots, in the simulation the world interface is simple as it is much easier to implement the symbolic sensor assumption in the simulated world than it is in case of real robots, where sensors do not deliver symbolic information, and where an additional layer of sensory data processing has to be defined to transform subsymbolic into symbolic information. See [MPT95] for a discussion of several definitions of interfaces between subsymbolic and symbolic data processing.

Perception. The internal model of perception of the forklift agents matches the grid-based representation of the environment: a forklift has a rectangular *range of perception*, i.e., an area within which it is able to perceive changes. In the examples of this work, we will assume that the agent's range of perception is *one square ahead*. Of course, the agent must observe its own state, e.g., the position of its gripper, whether it is holding a box or not, and if so, what box it is holding.

In the simulation of the loading dock, the agents obtain their perception from the simulation world: each time that either the range of perception of an agent changes (e.g., while moving around) or that anything changes within the agent's range of perception (e.g., another agent moving through it), the agent receives a record containing its current perception. This information is stored in the agent's *perception buffer*, which is part of the world interface.

In each cycle of the behavior-based layer (see Section 3.4.3), the agent knowledge base is updated with the new sensor information. In order to access sensor information from the world, the simulated forklift has a single sensor at its disposal, which is defined as an instance of the general class **sensor**:

```
class camera
super sensor
[...]
new(camera)(
  name: cam1
  range: symbol          /* returns a symbolic expression */
  value: nil )           /* initial value = nil */
```

At each point in time, the value of the camera sensor is a symbolic description of the square in front of the agent. It is a fact of the form

```
square(xcoord:x ycoord:y type:t objectlist:1)
```

giving the coordinate of the square, its type (i.e., ground, shelf, ramp, or wall), and a list of objects which are currently on the square (i.e., at most one forklift and/or at most one box). Forklifts and boxes are denoted by their names.

The task of sensory data processing in the simulated scenario is to translate the representation of the sensory data returned by the camera into the AKB knowledge representation. This means identifying relevant objects (squares, forklifts, and boxes) and updating the corresponding facts in the knowledge base. Object identification is done through the primary keys (i.e., x- and y-coordinates for squares, names for forklifts, and names plus types for boxes). A set of perception rules is applied to insert new facts into the knowledge base (e.g., if the object list contains a previously unknown agent, a corresponding fact is inserted) and to modify existing entries (e.g., if the object list contains a forklift and a box, a reference to the corresponding fact for the box is assigned to the carries attribute of the fact denoting the forklift).

Action. Forklifts can perform different actions to effect changes in their environment, e.g.:

- move to the next square;
- take a 90° turn to the left or to the right;
- pick up a box in front of them;
- put down a box that they are holding;
- raise and lower the gripper;
- open and close the gripper.

In the simulation, the execution of an action is initiated by a message to the simulation world. The simulation world maintains a set of integrity constraints that no action must violate. In the following, the individual actions are described.

Moving. The action of moving one square ahead is defined by creating an instance of the class actor defined in Section 3.3.2:

```
new(actor)(
  name: move-ahead
  params: nil               /* no input parameters */
  type: atomic )            /* treated as an atomic action */
```

The action is atomic, i.e., its execution either returns a success and changes the position of the agent to be one field in front relative to its old position, or it fails; in the latter case, it is assumed that the position of the robot is not changed.

The move-ahead action is axiomatized in the simulation world. Apart from the aforementioned effects of the action in case of success and failure, respectively, the following constraints are imposed on the execution of move-ahead:

1. move-ahead only succeeds if the destination field is of type ground, and if it is not occupied by any other object.
2. Successful execution of move-ahead only changes the x- or y-coordinates of the forklift; it does not modify other properties of the world (by this assumption, we solve the frame problem in our application).
3. It is not possible for agents to move through each other; i.e., constraint (1.) also holds if the field in front of the agent (e.g., f_1) is occupied by another agent f_2 who is facing f_1 but initiates a move-ahead action simultaneously to f_1.

Turning. The forklift is able to perform two actions that change its orientation; they are defined as follows:

```
class turn
super actor
attributes
  params: Dir = [left, right]   /* direction as input parameter */
  type: atomic                  /* treated as an atomic action */
[...]
new(turn)(
  name: turn_left
  params: left)

new(turn)(
  name: turn_right
  params: right)
```

Turning is a fairly uncritical action in that it has only one precondition: it is only possible if the gripper is up. This restriction is to avoid collisions with other objects or agents while turning; furthermore, it prevents the agent from trying to turn while exchanging a box with another agent. The effect of a turning action is that the orientation of the agent changes in the intuitively expected way. We also require that turning does not change any other properties of the world.

Manipulating boxes. There are two operations allowing the forklift agent to manipulate boxes: it may grasp a box and it may put down the box it is holding. The following are the class definitions for both operations:

```
class manipulator
super actor
attributes
  type: atomic                /* treated as an atomic action */
[...]
new(manipulator)(
  name: grasp-box
  params: Boxname=symbol)     /* box as input parameter */

new(manipulator)(
  name: put-box
  params: Boxname=symbol)     /* box as input parameter */
```

The preconditions of the action `grasp-box(b)` are the following:

- The agent must stand in front of a shelf or the ramp.
- The box b that the agent shall grasp, must stand on the shelf/ramp, and the agent must face b.
- The agent must not hold any other object, i.e., `carries = nil, hands=open`.
- The gripper must be in driving position, i.e., `gripper=up`.

Only in case all of these preconditions hold, the action succeeds; in case of failure, we assume that the execution of `grasp-box` effects no changes in the world. In case of success, the effects of the action are as follows:

- The corresponding shelf/ramp position is empty.
- The agent holds box b (`carries=$b, hands=closed`)[2].
- The gripper is in driving position (`gripper=up`).
- No other properties in the world are changed.

Preconditions and effects for the action `put-box` are defined analogously.

Primitive gripper operations. There are some finer-grained gripper actions that are needed to implement the exchange of boxes between forklift agent: open-hands, close-hands, gripper-up, and gripper-down. They are defined as special manipulator actions in the following way:

```
new(manipulator)(
  name: open-hands
  params: nil)      /* no input parameter */

new(manipulator)(
  name: close-hands
  params: nil)
```

---

[2] `$b` denotes the object identifier of the fact denoting box b (see Section 4.2.1).

```
new(manipulator)(
  name: gripper-up
  params: nil)

new(manipulator)(
  name: gripper-down
  params: nil)
```

Gripper-up and gripper-down change the position of the gripper arm between driving position (the gripper is positioned vertically) and gripping position (the gripper is positioned horizontally). Open-hands and close-hands
are self-explanatory. The abstract manipulator actions defined above can be
composed from the primitive gripper operations:

grasp-box = gripper-down; close-hands; gripper-up,

and

put-box = gripper-down; open-hands; gripper-up.

However, the axiomatization of the manipulator operations contains domain-
specific constraints that are not part of the axiomatization of the primitive
gripper operations, such as e.g., checking the object in the gripper before
grasping/storing it.

Communication. In order to communicate with other agents, the forklift
agents in the loading dock make use of the general communication facilities
defined in Section 3.3.2. Communication between agents is physically maintained by providing a low-level protocol based on TCP/IP stream sockets,
allowing agents to open and close connections to each other, and to transmit
and to receive arbitrary sequences of characters. The use of stream sockets ensures sequencing (i.e., the temporal order of messages is not modified during
transmission) and safety of communication. We use the basic communication
services of the MAGSY language [Fis93b] to implement the communication
subsystem of INTERRAP. The TCP/IP level is a very general level of communication; it enables INTERRAP agents to exchange information with any other
UNIX process, thus, embodying the idea of an *open system*.

The high-level language for communication between INTERRAP agents is
currently related to KQML [FF94] [MLF96], but considerably simpler and less
expressive. A message is represented by an identification, a sender, a recipient,
a reference number, a message type, and a content string (see also page 56).
An extension to a more expressive communication standard is subject to
future work.

4.2.3 Reaction and deliberation

Agents in the loading dock need to plan in order to perform their transportation tasks in a goal-directed manner. At the same time, they need to be recep-
tive to unforeseen events like threatening collisions with other agents. As the

situation may change dynamically, planning and execution must be closely interleaved. Thus, this domain allows us to study the reconciliation of reaction and deliberation. It is achieved by the interplay between the two lower control layers of the INTERRAP architecture, i.e., the behavior-based layer and the local planning layer. In the following, we present important operational aspects of both layers by examples taken from the loading dock: the interleaving of hierarchical planning and intelligent execution, the achievement of reactive behavior, and mechanisms for maintaining coherence between reaction and execution on the one hand, and planning and deliberation on the other.

Deliberation and intelligent execution. The local planning layer (see Section 3.5) enables the agent to work towards the achievement of its goals through means–ends analysis. For this purpose, a set of predefined plans are stored in a plan library (see Section 3.5.2). Forklift agents have only two top-level plans: to load a box onto the ramp, and to unload a box from the ramp and to store it on a shelf. Figure 4.3 shows a simplified excerpt of the plan library showing the top-level plan for the former goal. Each entry of

```
lpb-entry(load-ramp(R, B),
  [ do(fetch-box(B)), do(store-box(B, R))])
lpb-entry(fetch-box(B),
  [ Reg=kb←query((box-pos(B))), do(goto-region(Reg)),
    do(search(B))])
lpb-entry(fetch-box(B),
  [ pob(random-search(B))]). /* pattern of behavior */
lpb-entry(goto-region(Reg),
  [ L₀=kb←query(my-pos), L₁=kb←square-in-region(Reg),
    generate(dijkstra(L₀, L₁))]) /* sequence of goto-landmark's*/
lpb-entry(goto-landmark(L),
  [ pob(goto-landmark(L))])
...
```

Fig. 4.3. The local plan library

the plan library is a tuple consisting of two parts: an atomic index formula denoting the goal that is achieved by the plan, and the body, which is an $\mathcal{L}_0$–plan expression (see Definition 3.5.1). The plans are hierarchical: e.g., the top-level goal load-ramp(R, B), where R and B are variables denoting a specific ramp and box, respectively, is expanded into two conjunctive subgoals fetch-box(B) and store-box(B, R).

Plan expansion. Planning is initiated upon reception of an upward activation request from the behavior-based layer. This request is sent to the local planning layer by a message

$$lpl←request(activate(load-ramp(r_1, b_{23}))),$$

where a new plan stack is created and initialized with the top-level goal load-ramp. In each cycle, the top element of each plan stack is processed by the plan interpreter (see Section 3.5.3). To process a plan stack element means either to expand it by recursively refining it into subplans, or to send a message to a neighboring layer in case the top-level element is a primitive plan step (i.e., a leaf node in the hierarchical plan, see Figure 4.4). A special function is generate(F), which has a function F as its argument. An expression generate(F) is interpreted by (i) popping it from the stack, (ii) evaluating F, and (iii) pushing the result of F on the stack. In the example, the function dijkstra returns a set of goto-landmark plan steps, connecting the start landmark with the destination. The goal goto-landmark is achieved

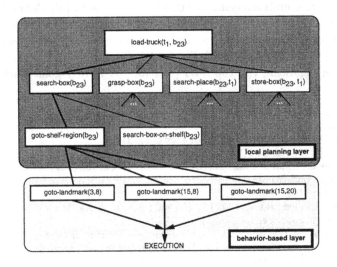

Fig. 4.4. An example for goal expansion

by a procedure PoB. In this case, a corresponding commitment message is sent to the behavior-based layer. After initiating the execution of the PoB, plan expansion for the plan stack of the top-level goal load-ramp suspends until the behavior-based layer reports the end of the execution. In case the execution has been successful, plan expansion continues with the next element on the plan stack. In case the plan stack is empty, the top-level goal is achieved, and an acknowledgment message

$$bbl \leftarrow inform(done(load\text{-}ramp(r_1, b_{23}), success))$$

is sent to the behavior-based layer.

Failure recognition. The planner has two possibilities to determine the outcome of executing a procedure: firstly, it can rely on the return value of the PoB, which is success or failure in the simplest case. Secondly, it can check whether the PoB has caused the expected effects by evaluating its post conditions: it accesses a symbolic description of the expected effects of a PoB and starts a corresponding query to the knowledge base. Whereas the former possibility allows a simpler design of the local planning layer, it enhances the complexity of the behavior-based layer and of the individual PoBs, as each PoB needs to thoroughly determine the reasons causing it to end. Thus, whereas INTERRAP offers both possibilities, we claim that it is reasonable to enable the planner to do in-depth reasoning about failure detection by giving it access to knowledge about expected and actual effects of PoB execution, and to restrict the failure detection capabilities of the behavior-based layer to a degree that prevents fatal outcomes by causing the corresponding PoB to stop and to take suitable action immediately. For instance, a PoB goto-landmark that cannot be achieved because the destination landmark is occupied by another agent should recognize that the destination is occupied, and it should be able to avoid crashing into the other agent. However, a detailed analysis of *why* the landmark cannot be reached is not that time-critical and should therefore be accomplished at the local planning layer.

Failure handling. If the execution of a procedure PoB fails, backtracking to alternative solutions may occur. Alternatives can be expressed either by disjunctions in the plan, or by alternative entries in the plan library. For instance, there are two entries for the goal fetch-box in Figure 4.3; therefore, the function expand (see page 84) returns two plans. The plan selection function (see page 89f) chooses one of these plans for execution; the other alternatives, however, are stored as choice points for replanning. In the current implementation of the loading dock, plans are selected top–down, i.e., according to their order in the plan library. Thus, in the above example, the planner first expands fetch-box into a subplan consisting (1) find out in what region it expects the box to be, (2) move into that region, and (3) search for the box there. This plan is conjunctive and contains no built-in alternatives; hence, if a failure occurs, plan execution fails, and the planner tries to achieve the goal by using the second alternative, which is a procedure PoB that—for the sake of simplicity of the example—makes the agent wander around randomly until it has found the box.

Intelligent execution. While failure handling and replanning are interesting for their own sake (see Section 6.2), one central hypothesis of our work is that at the local planning layer they should be *avoided* as far as possible rather than being *optimized*. Plans are specified at a layer that is general enough to make them *universal*[3], i.e., to assume that the top-level plan does not fail under normal circumstances. Thus, the second entry for the plan step fetch-box in Figure 4.3 has been defined for reasons of robustness and should be viewed as an emergency plan rather than a serious alternative to the former: only if the former plan fails for some unexpected reason, the agent should switch to this emergency behavior.

---

[3] The notion of *universal plans* was coined by Schoppers [Sch89].

The universality of plans is ensured by an *intelligent execution* mechanism at the behavior-based layer. The leaf nodes of plans are implemented by procedure PoBs that realize complex actions in the world. For example, one procedure occurring in the above example controls the movement of the agent between two landmarks. In the loading dock, the local planner solves the navigation task only inasmuch as it defines a set of plan steps at the level of landmarks connecting different regions (i.e., ramp, parking, hallway, or shelf regions). However, there are many different trajectories between two landmarks. At run-time, some of these trajectories will be infeasible, e.g., because squares may be occupied by other agents. The PoB goto-landmark (see Figure 4.5) uses a simple dead-reckoning navigation strategy: the agent tries to minimize the difference to the destination landmark in either possible direction.

A variant of Dijkstra's shortest path algorithm is used for the generation of goto-landmark sequences. However, if a specific square is occupied at run-time (i.e., while performing the PoB goto-landmark), explicit replanning is started at the local planning layer only if this square has been the destination square of the procedure call goto-landmark; otherwise, the behavior-based layer dynamically modifies its path between the two landmarks. This can be achieved by defining an exception for goto-landmark (see Section 3.4.2), which is compiled into a reactor PoB that monitors the exception while the goto-landmark is active, and which takes action once the exception condition occurs.

Reactive behavior. This is ensured by reactor PoBs, which can be defined either as special monitoring conditions for procedures or stand-alone, i.e., independent from other PoBs. An example for the latter is a reactor PoB treat-order that recognizes the arrival of a new transportation order, and that creates an upward activation request to the local planning layer. An example of the former is an exception that occurs in the above goto-landmark example if the square to which the PoB has decided to move next is occupied by an unexpected obstacle. This exception is recognized and handled by a reactor PoB dodge (see Figure 4.6), which is dynamically created from the description of goto-landmark, and which is monitored while goto-landmark is pursued.

In case the field in front of the agent is occupied, dodge becomes active. As it is a reactor, it has a higher priority than goto-landmark and it will be selected by the execution mechanism before goto-landmark. The body of dodge is simple: it selects an arbitrary direction, moves one square into this direction, and terminates itself, allowing goto-landmark to take control again. In many cases, this is sufficient to pass the obstacle (see Section 4.3 for a discussion of cases where it is not).

Clean-up tasks. In case of a failure, an agent needs to re-establish a consistent state that allows it to continue. For this purpose, special clean-up PoBs can be defined for procedures. To give an example of a clean-up task, we shall

```
( PoB
     name:                  goto-landmark
     type:                  procedure
     args:                  (DestX,DestY)
     succ-cond:             my-pos(DestX,DestY)
     exception-list:        [(return-status(move-ahead) = failure,
                                enable(dodge))]
     resources:             [motor]
     body:
     {
            while true do
                       activate(turn-to-free-dir(DestX,DestY));
                       ex(move-ahead);
            od
     } )
```

Fig. 4.5. The procedure PoB goto-landmark

```
( PoB
        name:              dodge
        type:              reactor
        act-cond:          object-in-front
        fail-cond:         (self←nb-steps > 3,
                               enable(recognize-blocking-conflict))
        resources:         [motor]
        body:
           {
             activate(step-aside);
           } )
```

Fig. 4.6. The reactor PoB dodge

leave the simulation scenario for a moment, and look at the implementation of the loading dock based on miniature robots. Here, the action of driving one square ahead cannot be regarded as being atomic: due to the short range of the robots infrared sensors, a situation like the one illustrated in Figure 4.7 can occur. Both robots *a* and *b* in the example decide to move to the next field. In the real world, however, the execution of this action is a continuous process rather than an atomic event. Thus, while moving, a robot may register that its destination is occupied. In this case, the move-ahead action has to be aborted; however, at this point in time, the agent is in an undefined state between two squares, and a clean-up task must be launched to reach a consistent state again.

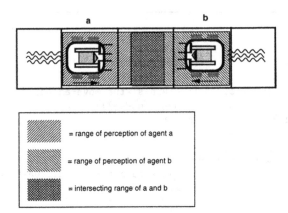

Fig. 4.7. Driving one square is not atomic in the real robot case

For this purpose, in the real-robot world, move-ahead is defined as a procedure rather than as an actoric primitive[4]. This is illustrated in Figure 4.8. Move-ahead concurrently activates two procedure PoBs drive, which makes the robot drive forward, and follow-line, which implements a control algorithm for moving along white lines painted on the floor of the loading dock by means of three infrared sensors. The aforementioned case where an obstacle is realized while moving is represented by an exception in the definition of the PoB drive. This exception is monitored by a reactor PoB avoid-collision, which causes the forklift to halt if it encounters an obstacle. Thus, whenever an obstacle occurs while the forklift is driving, first, the reactor avoid-collision becomes active bringing the robot to a halt and causing the PoB drive to fail. Failure of drive then causes the procedure move-ahead to fail. However, a clean-up PoB recover-position is associated with the failure of move-ahead. This makes the robot turn around and drive back to the center of the previous square.

Note that the desired ordering in the execution of the PoBs in the above example is ensured by the scheduling strategy of the behavior-based layer. The failure of drive is treated by activating the reactor avoid-collision before the failure of move-ahead is registered, as reactors have a higher priority than procedures. Moreover, child PoBs have a higher priority than their parents. Thus, while e.g., move-ahead and drive are active concurrently, the more specific PoB drive will be pursued, while its parent move-ahead remains suspended while the children are active.

---

[4] A parameterized move-ahead is used that allows the robots to move for a specified number of squares without stopping.

```
( PoB
    name:          move-ahead
    type:          procedure
    fail-cond:     [(object-in-front,activate(recover-position))]
    succ-cond      reached-new-field
    resources:     [lmotor, rmotor] /* left/right motor */
    body:
    {
                   activate([drive, follow-line]);
                   suspend;
    } )

( PoB
    name:          drive
    type:          procedure
    fail-cond:     [(object-in-front,enable(avoid-collision))]
    body:
    {
                   ex(set-speed([lmotor,rmotor],default-speed));
                   suspend;
    } )

( PoB
    name           avoid-collision
    type           reactor
    act-cond:      status(driving) ∧ object-in-front
    body:
                   { ex(set-speed([lmotor,rmotor],0)); } )
```

Fig. 4.8. Clean-up patterns of behavior

4.2.4 Coherence

A problem in architectures like INTERRAP where different independent control layers interact and make decisions is that of avoiding interferences of these layers and of achieving *coherence* between their control decisions. We draw a distinction between two types of coherence, i.e., *bottom-up coherence* and *top-down coherence*. Bottom-up coherence is the problem of finding an appropriate layer to deal with a given situation. Top-down coherence is the problem of avoiding harmful interactions between the longer-term goals of the agent (represented in the cooperative planning layer and the local planning layer) and its short-term reactive behavior (represented in the behavior-based layer).

Bottom-up coherence. The hierarchical structure of the control layers in INTERRAP allows to solve the problem of bottom-up coherence at design time. The agent designer needs to provide specifications of the competence of each layer. For each situation that does not match this competence description, control is shifted upward to the next higher layer. Other architectures, which do not provide this hierarchically restricted flow of control (e.g., Ferguson's TouringMachines, see Section 2.6) need to provide global rules for filtering the input to different layers in order to guarantee that control is exerted at the right level of abstraction. A concrete example of how the problem of bottom-up coherence is solved in INTERRAP is described in Section 4.3, where different types of conflicts are resolved by using mechanisms located at different control layers.

Top-down coherence. The problem of top-down coherence is more difficult to solve. In the following, we shall clarify it by means of some examples.

Example 1—docking behavior and collision avoidance. This example, which is illustrated in Figure 4.9, refers to the real-robot domain: robots have infrared sensors and a reactor PoB for collision avoidance that stops them if they get too close to an obstacle. One possible goal at the local planning layer is to go parking if no transportation order is to be carried out. The last step of the parking subplan is to actually dock to the home square. This, however, requires approaching the wall closely, which activates the collision avoidance PoB, leading to a deadlock in the worst case.

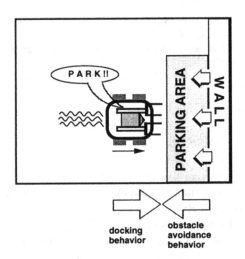

Fig. 4.9. Foreseeable interferences of PoBs (Example 1)

Example 2—navigation and dodging. A forklift that has the goal to navigate from one landmark to another may have to dodge quickly e.g., in order not to collide with another agent. This dodging is performed by the activation of a reactor PoB. However, after the dodging, the robot is at a different location. Moreover, as dodging can be done in different directions, the question is whether the PoB can act coherently with the agent's longer-term goal by, e.g., moving to the side which is closer to the goal landmark (see Figure 4.10).

Fig. 4.10. Longer-term goals vs emergency behavior (Example 2)

Example 3—commitments and obstacle avoidance. The problem of top-down coherence also occurs with regard to the reactive behavior and the cooperative goals to which the agent has committed itself, e.g., by accepting its role in a joint plan. E.g., a robot that has accepted to resolve a conflict in a corridor by moving out of the corridor and staying there until the agent that is locked in the corridor has moved out, might encounter an obstacle while moving out of the corridor (see Figure 4.11). When executing the obstacle avoidance PoB, the agent must not return into the corridor, as it has committed itself to keep out.

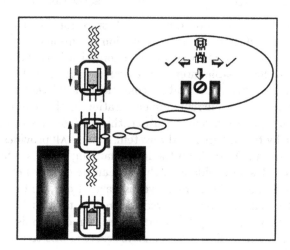

Fig. 4.11. Multiagent commitments restrict local decision-making (Example 3)

The three cases described above are examples for top-down coherence problems, as there are conflicts between the low-level behavior achieved by reactor PoBs and the agent's longer-term goals and commitments. However, each of them represents a different class of coherence problems. The first example describes a foreseeable interaction between two PoBs park and avoid-collision. It is foreseeable because each time the park behavior will attempt to move to the docking square, it will interfere with the collision avoidance behavior. The second example describes a dynamic interaction between a reactor PoB and a longer-term goal. Time and place of this interaction are not foreseeable. Moreover, the interaction is desired as the robot is actually supposed to dodge when it is faced with an obstacle. However, the question is how the execution of the reactor PoB can be done such that it fits the robot's longer-term goal to approach the destination landmark. The third example is different from the second as it not only restricts the execution of the reactor with respect to the local goals of the agent, but also forces it to respect commitments to other agents, whose violation has much more serious consequences than the violation of the agent's local goals, as it may require costly replanning of coordination processes.

The INTERRAP architecture provides various mechanisms to achieve top-down coherence by dealing with these different types of interferences. They are discussed in the following paragraphs.

Foreseeable interactions. Foreseeable interferences of two PoBs can be discovered in advance. A potential interference of two PoBs b_1 and b_2 occurs if

- at least one of b_1 and b_2 is a reactor PoB (we assume w.r.o.g. that b_2 is a reactor), and if
- the effect of the execution of b_1 implies the activation condition of b_2.

E.g., an effect of the park procedure in the first example is that the agent approaches the wall, which implies the activation condition of avoid-obstacle, i.e., that an obstacle is in the way. Given axiomatizations of these conditions, the detection of foreseeable interferences can be automated by allowing the planner to analyze the effects and preconditions of different PoBs. However, the planner used in the current implementation of the loading dock is not able to do this type of reasoning. Rather, this analysis has to be carried out at design time by the designer of the system. INTERRAP provides the possibility to disable specific PoBs from the local planning layer (see Section 3.4.4). That way, the collision avoidance behavior can be disabled explicitly before the last step of the docking maneuver is carried out, and can be enabled again after the goal has been achieved. Currently, this disabling step has to be built into the plan library by the designer. A more sophisticated planner could explicitly plan to disable and enable reactor PoBs to avoid foreseeable interferences.

Interferences through emergency behavior. This sort of interferences is more difficult to handle. We even claim that it is impossible to handle in general for a resource-bounded agent. In the second example (see Figure 4.10), guaranteeing coherence between the execution of the dodging PoB and the longer-term goal of navigating to an appropriate landmark would require the PoB to take the top-level goal into account when deciding which way to dodge. This, however, actually means to plan into which direction to move to avoid an obstacle, which in turn requires a certain amount of computation. However, Example 2 describes an emergency situation, where the time to react may be bounded, e.g., by the distance of the approaching agent. Thus, in some cases, the agent will just have to move into some direction to avoid fatal consequences, no matter whether this decision coheres with its longer-term goals. Therefore, INTERRAP does not provide techniques that guarantee coherence between the execution of reactors and the goals of the agent in all cases. Reactors may fire and lead to a new world state that is, from the perspective of the local planning layer, less advantageous than the previous state. However, since the planner has access to the world model, it can discover these changes, and adjust plans appropriately.

Interferences with multiagent commitments. The avoidance of interferences of reactor PoBs with higher-level goals that are grounded in commitments made to other agents (see Example 3) often does not require disabling a whole class of PoBs, but only restricting their execution. It should be clear from the discussion of emergency behavior that this is not possible in all cases. However, INTERRAP provides methods for making these restrictions on the behavior of PoBs in cases where it is possible. For example, agent a_1 in Figure 4.11 clearly should dodge by moving around the obstacle. However, while dodging, it should not return into the corridor if possible, in order not to break its commitment.

In order to allow such context-sensitive restrictions of the execution of PoBs, INTERRAP offers the `inhibit` command (see Section 3.4.2). In the third example (see Figure 4.11), after leaving the corridor, the PoB `move-ahead`[5] of agent a_1 can be restricted from the local planning layer by sending the command

    bbl←inhibit(move-ahead,in-front(4,3)),

which adds the expression `in-front(4,3)` to the termination condition of the PoB `move-ahead`. Thus, whenever the agent is facing the entrance field to the corridor, the execution of the PoB `move-ahead` is prohibited. Thus, the square is not entered and the agent will take another way around the obstacle. If the execution of the joint plan has proceeded such that agent a_2 has acknowledged that it has left the corridor, the inhibition can be lifted by

---

[5] Note that in the real-robot implementation, `move-ahead` is actually modeled as a PoB, whereas it is a world interface primitive in the simulation (see also Footnote 4 on page 142).

sending a bbl←permit(move-ahead,in-front(4,3)) message that retracts the restriction from the termination condition of the PoB.

4.3 Agent Interaction in the Loading Dock

Our objective in this section is to define different types of interactions among agents in the loading dock, and to study local as well as communication-based interaction mechanisms. We shall start by clarifying some basic assumptions and restrictions.

4.3.1 Basic assumptions

In the literature (see e.g., [Syc87]), often a distinction is made between two types of interactions: positive interaction or *collaboration*, where agents work together to achieve certain goals, and negative interaction or *conflict*, where the actions, goals, or plans of different agents interfere and have to be coordinated. Throughout this work, we shall concentrate on the latter type of interaction: we describe different types of conflicts that occur between the forklifts, and we show how INTERRAP supports the design of mechanisms for conflict recognition and conflict resolution. We focus on a specific type of conflict, i.e., on blocking conflicts between agents (see below).

Throughout this work, we make the following assumptions as regards the decision-making behavior of agents[6]:

- **Two-agent conflicts:** we restrict our considerations to conflicts between two agents. However, in Chapter 5, the performance of different strategies for the n–agent case is experimentally evaluated. Aspects of conflict resolution among more than two agents are also discussed in Section 6.2.
- **Isolated encounters:** the cooperative behavior of agents does not depend on previous encounters.
- **Truthfulness:** agents are assumed not to lie, and to reveal their preferences, goals, and plans truthfully. We also assume that agents are cooperative insofar as they do not deny participation in a cooperative conflict resolution.
- **Utility maximizers:** agents try to maximize their utility in interactions; by defining their utility function, the designer of the system can decide whether they are local or global utility maximizers.
- **Identical utility functions:** we assume that all agents use the same local utility function. This enables agents to estimate the utilities of specific

---

[6] See [Ros85] for a discussion of possible assumptions in the design of interacting agents.

plans for other agents by applying their own utility functions to other agents' roles in a plan[7].

- **Complete knowledge about agent capabilities:** the capabilities of each agent, i.e., the plans it can execute are commonly known. This is a prerequisite for central generation of joint plans by individual agents. For a decentralized approach to the generation of multiagent plans, see [DL89].
- **Incomplete world knowledge:** we do not assume that agents have complete descriptions of the conflict situations they are in. Especially, negotiated plans can turn out not to be executable.
- **No partial goal satisfaction:** We do not consider partial satisfaction of goals, as presented by [ZR91], i.e., the cooperative conflict resolution process either fully succeeds or fails.

Throughout this section, we proceed as follows: first, examples for blocking conflicts in the loading dock are given. Then, the framework for recognizing and classifying conflicts at different control layers is described. Based on this classification, we then look in more detail at two general mechanisms for conflict resolution: local, behavior-based mechanisms as well as cooperative mechanisms, the latter involving coordination by communication.

4.3.2 Blocking conflicts

We draw a distinction between two classes of conflicts, i.e., *physical conflicts* and *goal conflicts*. Physical conflicts can be recognized by agents through perception. They result from resource access conflicts or from situations violating the domain constraints, e.g.,

1. an agent wants to move to a square that is occupied by another agent;
2. two agents that face each other try to "swap places";
3. two agents try to move to the same square at a time;
4. two agents try to pick up the same box simultaneously;
5. two agents try to store two boxes onto the same storage place.

The former three types of physical conflicts are called *blocking conflicts*. They shall be investigated in this section. We will focus on the first two cases; the third case can be reduced to the first by assuming that the movements of the agents are not made simultaneously, but can be serialized. In that case, the move-ahead action of one agent succeeds, and the other—the losing—agent is then faced with situation (1).

Physical conflicts have their reason in goal conflicts. The latter is a higher-level notion describing the interference among the goals of different agents. Conflicts between goals manifest themselves in conflicts between plans or

---

[7] Because of the truthfulness assumption, agents could obtain the local utilities of a specific plan by simply asking other agents. Hence, this assumption is not strictly necessary, but rather simplifies the negotiation process.

actions that are performed to achieve these goals. As opposed to other approaches to conflict resolution, we do not assume that there is a central global instance that is able to predict and to resolve conflicts. Rather, we assume that conflicts are recognized by the individual agents, and that there is no central mediator available (see [Syc87] for an example of conflict resolution through the help of a mediator) to resolve them. Due to the limited sensoric capabilities of the forklift, potential conflicts cannot be recognized a long time ahead, but only as soon as they become physical in a sense that the next action to be executed would fail. The task for the agents is now to find out the goal conflict that underlies the physical conflict.

In Figure 4.12, we illustrate the notion of conflicts with a few intuitive examples, the first of which (Figure (4.12.a)) describes a blocking conflict between two agents a_1, and a_2 in a wide hallway. a_1 is on its way back to the parking zone after storing box b_1 on the corresponding shelf, while a_2 has the goal to load box b_2 onto the ramp. Whereas both agents might get involved in a negotiation about which of them should dodge into which direction, this is not exactly what one would expect in this situation. Rather, the agents should move around each other without any explicit synchronization, e.g., by selecting a direction randomly, or by obeying existing conventions.

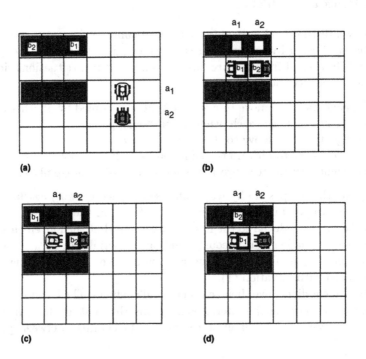

Fig. 4.12. Blocking conflicts in the loading dock

The next three figures show examples, where such local mechanisms do not seem that promising, as the agents are stuck in a narrow corridor. In Figure (4.12.b), agent a_1 is leaving the corridor to load box b_1 onto the ramp, while agent a_2 has the goal of storing box b_2 on the shelf to its right. An intuitive solution in this case would be that a_2 moves back and allows a_1 to leave the corridor. Figure (4.12.c) shows a similar situation. This time, however, a_1 has stored box b_1 onto the shelf and now has the goal of parking. Bearing in mind that agents are cooperative, there are at least two intuitive solutions for this conflict: first, assuming that a_1 is able to check whether there is a free space on the shelf to its left, it could adopt a_2's goal, receive the box b_2 from a_2 by a coordinated sequence of actions, and store it onto the shelf. Second, it could move one square back in order to allow a_2 to carry out its order by itself. Which alternative to choose might depend on criteria like the required number of actions or the costs of performing different actions. In Figure (4.12.d), a_1 is already holding box b_1 and intends to store it onto the ramp, while a_2 is searching for box b_2. A clever solution for this case is that both agents exchange their goals: a_1 hands b_1 to a_2, which loads it on the ramp, while a_1 itself adopts a_2's original goal to pick up b_2. Note that the situation illustrated in Figure (4.12.c) is a special case of that shown in Figure (4.12.d) as a_1 does not have a goal that a_2 could adopt in the former case.

Thus, in the loading dock, there is a variety of blocking situations that the agents have to deal with. Different situations have to be recognized, classified, and the agents have to choose among various mechanisms for resolving these conflicts.

4.3.3 Conflict recognition

In Section 3.2, we defined situations as structures consisting of three parts: the external context, the mental context, and the social context. The recognition of conflict situations is done incrementally at the different control layers of INTERRAP.

Recognition of physical conflicts—the external context. The external context is recognized in the agent's world model by reactor PoBs. For example, the external context of the conflict situation illustrated in Figure (4.12.a) is given by the definition of the PoB `recognize-blocking` shown in Figure 4.13. The activation condition of that PoB describes the external context of a blocking situation, i.e., the agent (denoted by `self`) faces another agent in front of it. `Ahead` and `opposed` are domain predicates that are interpreted over the world model. Note that when the conflict recognition PoB becomes active the agent has stopped moving, as the reactor PoB `avoid-collision` (see Figure 4.8) should have fired before to avoid crashing into the other agent. Thus, as opposed to a threatening collision, a blocking conflict is a static situation. After recognizing the conflict, the PoB waits for a random period of time. In case the other agent moves away during this time, the termination condition of `recognize-blocking` becomes true and the PoB is terminated. Otherwise, a PoB `resolve-conflict` is activated (see also Section 4.3.4).

```
( PoB
   name:      recognize-blocking
   type:      reactor
   act-cond: agent(name:self pos:(X₁,Y₁,O₁)) ∧
             agent(name:A pos:(X₂,Y₂,O₂)) ∧
             ahead((X₁,Y₁,O₁),(X₂,Y₂)) ∧
             opposed(O₁,O₂) ∧
             pob(name:move-ahead status:active)
   succ-cond:ahead((X₁,Y₁,O₁),(X₂, Y₂)) ∧
             square (xcoord:X₂ ycoord:Y₂ status:nil)
   body:
     {
       wait(rnd-time);
       activate(resolve-blocking-conflict
                     self (X₁,Y₁,O₁) A (X₂,Y₂,O₂))
     } )
```

Fig. 4.13. Definition of the PoB `recognize-blocking`

Recognition of goal conflicts. Using the PoB `recognize-blocking`, an agent can recognize the external context of a conflict situation by analyzing information about the environment. Then, behavior-based conflict resolution methods can be applied, and conflicts like that shown in Figure (4.12.a) can be effectively resolved. However, dealing with conflicts like those shown in Figures (4.12.b–4.12.d) requires the recognition of the goal conflict underlying the physical conflict, which is detected using world model information solely. Doing so, however, exceeds the reasoning capabilities of the behavior-based layer, and requires the activation of the higher control layers.

In the loading dock, blocking conflicts are classified according to the region in which they occur: we can distinguish between blockings in a shelf region, in the ramp region, and in a hallway region. The two latter conflict types are resolved locally; if a blocking in a shelf region is recognized, control is shifted upwards through the INTERRAP control hierarchy, where the conflict recognition is completed. At the behavior-based layer, conflicts in shelf corridors are recognized by the reactor PoB shown in Figure 4.14. It is a specialization of `recognize-blocking`, as the activation condition of the former subsumes that of the latter. The RETE pattern matching algorithm [For82]) automatically supports the specificity criterion when selecting rules from the conflict set. I.e., more specific rules have a higher priority than less specific ones. The mechanism for PoB selection provided by the current version of

INTERRAP does not automatically prefer PoBs with more specific preconditions; thus, either the agent designer has to explicitly assign a higher priority to the PoB `recognize-blocking-in-shelf`, or add conditions to make the activation conditions of the two PoBs disjoint. A selection algorithm that uses the RETE-algorithm and that supports the selection criteria of recency, specificity, and refractoriness is currently being developed.

```
( PoB
    name:       recognize-blocking-in-shelf
    type:       reactor
    act-cond:   agent(name:self pos:(X₁,Y₁,O₁)) ∧
                agent(name:A pos:(X₂,Y₂,O₂)) ∧
                ahead((X₁,Y₁,O₁),(X₂,Y₂)) ∧
                opposed(O₁,O₂) ∧
                pob(name:move-ahead status:active) ∧
                region(xcoord: X₁ ycoord: X₂ type: shelf)
    succ-cond:  ahead((X₁,Y₁,O₁) (X₂,Y₂)) ∧
                square (xcoord:X₂ ycoord:Y₂ status:nil)
    body:
      {
        wait(rnd-time);
        call(lpl,resolve-conflict(blocking-in-shelf,self,A))
      } )
```

Fig. 4.14. The PoB `recognize-blocking-in-shelf`

Recognition of the mental context. If a blocking conflict in a shelf region is detected, the local planning layer is activated; in addition to world model information, this layer has access to the agent's mental model, i.e., its current local goals and plans. There are two basic possibilities of handling blocking conflicts at this layer, i.e., either:

- to abolish the local goal that is blocked by the conflict, or
- to reach the goal by resolving the conflict by cooperation.

For example, if an agent observes that another agent blocks its way, it might decide to continue its search for a free space on the other side of the shelf, and thus abolish its original goal to move into the corridor. Alternatively, it could try to cooperate with the agent blocking its way to resolve the conflict. We believe that the former alternative is more important for applications where agents are not benevolent and not willing to reveal their goals, plans, and preferences (see [Tam95] [TR96] for examples of agent modeling in adversarial domains). We restrict the role of the local planning layer in blocking conflicts to determining the local goal that is affected by the blocking and passing the activation request enhanced with goal information to the cooperative planning layer. For this purpose, the local plan library contains the following entry:

```
(resolve-conflict(blocking-in-shelf,self,A),
 [ Goal=self←get-affected-goal(blocking-in-shelf),
   Path=self←get-plan-path(Goal),
   call(cpl,resolve-conflict(blocking-in-shelf,self,A,Goal,Path))])
```

The definition of the function get-affected-goal, which determines the goals that are affected by a specific conflict situation, is non-trivial in general. We shall restrict ourselves to its solution in the specific application: conflicts are always related to the execution of a certain plan, i.e., the local planning layer processes certain *plan stacks* (see Section 3.5.2)[8]. Goals that are currently pursued (i.e., for which there are currently active procedure PoBs) are the top elements of the plan stacks. Thus, the general problem is reduced to relating the blocking conflict to one of these top-level elements. In the loading dock, at any point in time, at most one navigation procedure (goto-landmark, leave-corridor) is pursued. Therefore, the assignment of conflicts to local goals is accomplished easily in our application.

The current navigation goal that is affected by the blocking situation provides access to the top-level goal: in Figure (4.12.b), agent a_1's procedure PoB, the execution of which caused the conflict, is to move out of the corridor (leave-corridor); the top-level goal, i.e., the root of the current plan stack is load-ramp(r_1, b_1).

The method get-plan-path(g) returns the complete path $p_1, \ldots, p_k$ where p_1 is the top-level node of the plan stack containing $p_k \equiv g$ as a leaf node. This goal path is sent to the cooperative planning layer as an additional parameter of the upward activation request resolve-conflict. It provides information about subgoal relationships between different goals of the agent. For example, the goal path for the above example is

```
[load-ramp(r₁,b₁),
 store-box(b₁,r₁),
 leave-corridor ]
```

Thus, the goal leave-corridor that has caused the conflict is a subgoal of store-box, which is a subgoal of the top-level goal load-ramp. Subgoal relationships can be used at the cooperative planning layer to find the appropriate level of goal interaction, and thus to resolve blocking conflicts in different ways (see Section 4.3.5).

Recognition of the social context. If the cooperative planning layer receives a request resolve-conflict(...) from the local planning layer, the description of the situation is completed by providing information about

---

[8] Of course, the agent may have no goal to achieve, i.e., its plan stack may be empty. In this case, however, it will not actively recognize a conflict, and only become involved in a conflict resolution process upon request by another agent.

the goals of the other agent (see Section 3.6. This is done by communicating goal paths or parts of goal paths. For this purpose, a QUERY protocol is used with $R = \{info\text{-}seeker, info\text{-}provider\}$, $P = (K, \{\pi_s, \pi_p\})$, and $K = \{start, \text{REQUEST}, \text{INFORM}, noop, done, fail\}$, and the protocols π_s for the information seeker and π_p for the information provider being as follows:

$\pi_s(start(r)) = \{\text{REQUEST}(r)\}$
$\pi_p(\text{REQUEST}(r)) = \{noop, \text{INFORM}(i)\}$
$\pi_s(\text{INFORM}(i) = \{done, fail\}$
$\pi_s(noop) = \{fail\}$

Figure 4.15 graphically illustrates the query protocol. It has two sources of

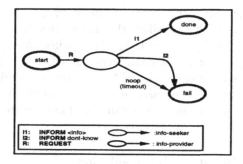

Fig. 4.15. A query protocol

indeterminism: firstly, upon receiving a request, the information provider has the choice of either sending an INFORM message containing the requested information, or doing nothing. The latter decision results in a timeout at the requesting agent, which causes protocol execution to fail. Secondly, even if the requesting agent receives an answer, it is still possible that the protocol fails if the answer is not appropriate. This is reflected in the strategies for both roles that are defined by the functions σ_s and σ_p for the information seeker and information provider, respectively:

$\sigma_s(start(req), \_, \_) =$
$\qquad \{(\text{REQUEST}(req), \_)\}$
$\sigma_p(\text{REQUEST}(req), \_, \_) =$
$\qquad \textbf{if } willing\text{-}to\text{-}tell(req) \textbf{ then}$
$\qquad\qquad info = retrieve(req);$
$\qquad\qquad (\text{INFORM}(info), \_)$
$\qquad \textbf{else } noop\}$
$\sigma_s(\text{INFORM}(info), \_, \_) =$
$\qquad \{\textbf{if } info = dont\text{-}know \textbf{ then } (fail, \_) \textbf{ else } (done, \_)\}$
$\sigma_p(noop, \_, \_) = (fail, \_)$

Note that the variables denoting the negotiation set and the agents' local utility functions are not used in this version of the protocol, as there is no negotiation in the strict sense (see also page 103).

Based on the goal information obtained that way, the agent decides whether there is a real goal conflict between the agents, and, if this is the case, initiates the negotiation process (see page 99ff). During negotiation, the agents generate a set of candidate plans and agree on a plan to choose by negotiation. The general negotiation process was described in Section 3.6.2. In Section 4.3.5, we shall give an example of its implementation in the loading dock domain.

Thus, the result of conflict recognition is a classification of the conflict by means of three parts, the external, mental, and social context. Each of the INTERRAP control layers can be involved in the conflict recognition process. However, participation of the local planning layer and the cooperative planning layer is optional. Thus, conflicts that can be recognized and handled satisfactorily by the behavior-based layer, are handled there without initiating planning and cooperation. The competence-based control flow in INTERRAP supports a flexible assignment of problems to the appropriate layers. In the remainder of this section, two classes of conflict resolution mechanisms will be examined in more detail: local, behavior-based mechanisms and cooperative mechanisms.

4.3.4 Behavior-based conflict resolution

In a physical environment where agents have to react to unforeseen events in real-time and where communication is often expensive and unreliable, the ability of dealing with certain conflict situations locally, i.e., without explicit synchronization through communication, is vital. In the literature, different approaches for interaction without communication have been proposed, including regulation by global *social rules* [ST92], recognition of other agents' plans [Kau90] and goals [Tam95], synchronization via low-level signaling (comparable to horns and winkers in car traffic) [CD96], and game-theoretic analyses [Ros85].

The local, behavior-based mechanisms that we describe in the following are considerably simpler than full-fledged plan and goal recognition, and they require neither the existence of global rules or conventions nor specific signaling techniques. Rather, they are robust and computationally cheap local algorithms. However, these algorithms are symmetrical insofar as they are used by multiple robots at the same time. Thus, avoiding deadlock situations that occur because of this symmetry is a central design problem that has to be considered. In the following, we propose two different approaches to behavior-based resolution of blocking conflicts. The first is a mechanism of explicit conflict resolution by means of special-purpose reactor PoBs; the second approach aims at incorporating conflict resolution abilities into the agent's navigation procedures at design time.

Handling conflicts by using reactor PoBs. In Section 4.2.3, we have shown how reactor PoBs can be used to recognize unforeseen situations. E.g., in the definition of the procedure PoB `goto-landmark` (see Figure 4.5), the occurrence of a blocking situation is monitored by an exception. If `goto-landmark` is activated by the local planning layer and selected for execution in the control cycle of the behavior-based layer, a reactor PoB is created and activated for each monitoring condition. Thus, if an obstacle is discovered during the execution of `goto-landmark`, a dodging behavior (see Figure 4.6) is enabled, which becomes active if the obstacle does not vanish, and which tries to move around the obstacle by calling a procedure PoB `step-aside`. The failure condition of `dodge` becomes true if the obstacle is not passed after a maximal number of trials. In this case, the conflict recognition PoB `recognize-blocking` (see Figure 4.13) is enabled, which becomes active if it discovers a blocking conflict with another agent, and which activates the procedure PoB `resolve-blocking-conflict` shown in Figure 4.16.

```
( PoB
   name:      resolve-blocking-conflict
   type:      procedure
   args:      [A₁,(X₁,Y₁,O₁),A₂,(X₂,Y₂,O₂)]
   succ-cond: agent(name:self pos:(X₃,Y₃,O₁)) ∧
              ahead((X₃,Y₃,O₁),(X₄,Y₄)) ∧
              square (xcoord:X₄ ycoord:Y₄ status:nil)
   fail-cond: (agent(name:self pos:(X₅,Y₅,O₁)) ∧
              ahead((X₅,Y₅,O₁),(X₆,Y₆)) ∧
              square (xcoord:X₆ ycoord:Y₆ status:A₂) ∧
              self←nb-steps > 10, /* end left-hand side */
              /* then shift control to local planning layer */
              lpl←request(activate(resolve-conflict(blocking,
                   self,A₂,unknown-type)))) /* end right-hand side */
   body:
     {
       dodging-dir=random-select([left,right]);
       if dir=left then
               ex(turn-left); ex(move-ahead); ex(turn-right);
       else
               ex(turn-right); ex(move-ahead); ex(turn-left);
       fi
     } )
```

Fig. 4.16. The PoB `resolve-blocking-conflict`

The procedure `resolve-blocking-conflict` randomly selects a direction and attempts to dodge to that direction, e.g., to the right. It remains active until the agent finds an unoccupied square in the direction into which it was moving originally (given by the argument o_1 in the above example). In that case, it ends and the conflict is resolved; otherwise, conflict resolution starts again.

Thus, blocking conflicts can be recognized and solved by defining special-purpose conflict resolution reactors, which in turn activate appropriate procedures. The reaction to blocking situations can be increasing stepwise, ranging from simple local dodging behaviors that work no matter whether the obstacle is an agent or an arbitrary object, over distinct blocking resolution PoBs to upward activation requests to the planner if the conflict cannot be resolved by the PoBs.

Implicit conflict handling: probabilistic decision functions. A different approach to the behavior-based resolution of blocking conflicts aims at incorporating domain knowledge into the agent's navigation routines and to treat blocking situations implicitly from within these routines. The main idea of this approach is the following: navigation routines such as `goto-landmark` (see Figures 4.4 and 4.5), which are activated by the local planning layer to reach destination landmarks, are implemented as *probabilistic decision functions*, i.e., as functions that, at each point in time, select the next action from the set of possible actions based on a probability distribution; it implements a hill-climbing strategy: the agent moves to the direction of its goal with a higher probability; however, locally suboptimal decisions are allowed in order to avoid the problem of local maxima. This technique is similar to the potential field navigation method used in robotics [Lat92]. However, it is a simplification of that method as the computation of the potential field also takes repulsing forces into account.

In a navigation procedure, the alternative actions are restricted: basically, the agent can move to any of the four squares surrounding it. In this approach, a blocking situation is interpreted by the agent as the corresponding square being occupied. If the agent notices that the square is occupied before it commits itself to an action, it simply deletes the action of moving to this square from the set of alternatives. Otherwise, it attempts to execute this action, fails, and tries an alternative.

The advantage of this approach is that there is no need for an explicit conflict recognition nor for conflict resolution at runtime. This is possible by modeling the navigation procedures appropriately at design time. In this section, we define the notion of probabilistic decision functions used to model this form of implicit conflict handling. In Section 5.2, we analyze the behavior of a multiagent system in which agents use probabilistic decision functions to perform navigation tasks.

A navigation procedure for the loading dock domain:. Figure 4.17 shows the procedure PoB `goto-landmark-pdf`. The procedure is described by a computation cycle. In each loop of the cycle, the current location of the agent is updated. Then, the set of alternative moves is computed. `Get-free-dirs` returns a set of directions `Dirs` $\subseteq$ `{n,e,s,w}` into which the agent believes to be able to move. An agent believes that it can move into a direction if it

```
( PoB
    name:                    goto-landmark-pdf
    type:                    procedure
    args:                    (X_d,Y_d)
    succ-cond:               Curr-pos=(X,Y,0) ∧ X=X_d ∧ Y=Y_d
    body:
    {
    while true do
        (X,Y,0)=kb←query(Curr-pos); /* retrieve current position */
        Dirs=kb←query(get-free-dirs(X,Y,0)); /* directions to go? */
        A= { moveto(D)|D ∈ Dirs} /* compute alternatives */
        Act = F((X,Y,0),A,(X_d,Y_d)); /* commitment to next action
                            using decision function F */
        activate(Act); /* activate PoB */
    od )
```

Fig. 4.17. A navigation procedure

believes that the neighboring square in that direction is of type ground and that its status is equal to nil, i.e., it is not occupied. The moveto action is implemented by a procedure PoB that performs the necessary turn and move-ahead actions.

The function $\mathcal{F}(s, A, g)$ is a decision function that selects one from the set A of possible moves, given the current position s and the destination g. In navigation the agent should select an action that is likely to approach its destination. For this purpose, $\mathcal{F} = \mathcal{F}^f$ is parameterized by a probability distribution f over the possible actions. In the loading dock, f is based on the notion of *quadrants* (see Figure 4.18): an agent divides the squares in the loading dock into four disjoint areas (quadrants) relative to its current position. Two predicates, same-quadrant and neighbor-quadrant, are used to relate different landmarks to each other. In Figure 4.18, we have

same-quadrant((x_1,y_1),(x_2,y_2))

and

neighbor-quadrant((x_1,y_1),(x_3,y_3)).

An example for a probability distribution function f is:

$$f(s, moveto(dir), goto-landmark(l)) = \begin{cases} 0.75 & : \quad same\_quadrant(sq(s,dir),l) \\ 0.1 & : \quad neighbor\_quadrant(sq(s,dir),l) \\ 0.05 & : \quad otherwise, \end{cases}$$

where $sq(s, dir)$ denotes the square neighboring s in direction dir. $\mathcal{F}^f$ is called a probabilistic decision function.

Probabilistic decision functions. In the following, let $\mathcal{S}$ be a non-empty set of world states (corresponding to a state of the agent's world model), and let $\mathcal{A}$ be a set of actions the agent may perform (either modeled by procedure

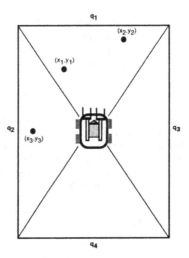

Fig. 4.18. Representation of quadrants

PoBs or by world interface primitives). The execution of actions causes state transitions. $\mathcal{G} \subseteq \mathcal{S}$ is a set of goals. In the context of navigation, goals describe specific landmarks the agent intends to reach. A probabilistic decision function is defined as follows:

Definition 4.3.1 (Probabilistic Decision Function (PDF)). *Let $\mathcal{S}$ be a set of world states. Let A be a non-empty set of actions, $\mathcal{G} \subseteq \mathcal{S}$ a set of goals; let $f(s, A, g) \in [0, 1]$ be a probability distribution on A given state $s \in \mathcal{S}$, and goal $g \in \mathcal{G}$. Then a PDF is a function $\mathcal{F}^f(s, A, g) = a_i$ with probability $f(s, a_i, g)$ for each $a_i \in A$. We omit the superscript f for $\mathcal{F}$ in cases it is obvious.*

The function f describes a probability distribution over the actions an agent may perform given a specific state of the world (i.e., its current position) and a specific goal state (i.e., its destination landmark).

A special case of probabilistic decision functions are those functions where f is uniformly distributed:

Definition 4.3.2 (Uniform Decision Function (UDF)). *A PDF $\mathcal{F}_u \equiv \mathcal{F}_u^f$ is an UDF iff $f(s, a, g) = \frac{1}{|A|}$ for all $a \in A$ and for all s, g in Definition 4.3.1.*

Using a uniform decision function in the navigation procedure of an agent implements a random walker: at each point in time, the agent selects its next action randomly. In a finite grid, navigation using a uniform decision function has the following properties:

Theorem 4.3.1. *Let $\mathcal{F}_u$ be an UDF, let $\mathcal{A} \equiv \{\text{moveto}(d)|d \in \{n,e,s,w\}\}$ be the set of alternatives. Let L be a finite grid of size $n \times m$, let (X_i, Y_i) denote an arbitrary square in L. Then:*

1. *An agent using $\mathcal{F}_u$ as a decision function will reach each square (X, Y) that is reachable from (X_i, Y_i) infinitely often.*
2. *For each $(X, Y) \neq (X_i, Y_i)$ in a nondeterministic scenario, there is no finite upper bound on the maximal number of steps required to reach (X, Y) for the first time.*

Given a set of agents that locally select their navigation actions by using PDFs, and assuming that these agents operate in a shared environment like the loading dock, we shall address the question as to the global performance of the system as a whole in Section 5.2: there, a probabilistic model is defined that allows to predict the behavior of a multiagent system where each agent performs its navigation tasks by using probabilistic decision functions.

4.3.5 Cooperative conflict resolution

The scope of behavior-based conflict resolution methods is restricted through the local perspective and the short look-ahead of these mechanisms (see also Chapter 5). In this section, we describe an approach to conflict resolution through cooperative planning based on the recognition of *goal conflicts*. If a goal conflict has been recognized at the cooperative planning layer, a negotiation process (see Section 3.6.2) is carried out by the agents to determine a joint plan to resolve the conflict cooperatively.

Joint plans. Figures 4.19–4.21 show excerpts of the joint plan library of the forklift agent containing different plans for resolving the blocking conflicts shown in Figures (4.12.b–4.12.d). The plans are written in the plan language $\mathcal{L}_2$ (see Definition 3.6.8).

```
( name:  blocking1
  roles: [F:FREE L:LOCKED]
  sit:   pos(F)=(X₁,Y₁,O₁) ∧ pos(L)=(X₂,Y₂,O₂) ∧
         between((X₂,Y₂),(X₃,Y₃),(X₁,Y₁,O₁))
  res:   blocked(F,goto-landmark(X₃,Y₃)) ∧ blocked(L,leave-corridor)
  ach:   achieves(F,goto-landmark(X₃,Y₃)) ∧ achieves(L,leave-corridor)
  body:  ([ (F,    [leave-corridor, goto-landmark(X₃,Y₃)]),
           (L,    [leave-corridor]) ],
         [(F,1)<(L,1),(L,1)<(F,2)]) /* precedence constraints */ )
```

Fig. 4.19. Joint plan for the resolution of blocking conflicts

```
( name:   blocking2a
  roles:  [F:FREE L:LOCKED]
  sit:    pos(F)=(X₁,Y₁,O₁) ∧ pos(L)=(X₂,Y₂,O₂) ∧
          carries-box(F, B₁) ∧ ¬carries-box(L,_) ∧
          between((X₂,Y₂),(X₃,Y₃),(X₁,Y₁,O₁)) ∧
          free-space-on-shelf(S,beneath(X₂,Y₂))
  res:    blocked(L,leave-corridor) ∧
          blocked(F,goto-landmark(X₃,Y₃))
  ach:    achieves(F,unload-ramp(R,B₁)) ∧ achieves(L, nil)
  body:   ([ (F, [goto-landmark(X₂,Y₂),turnto-dir(shelf(T)),
                   put-box(B₁)]),
             (L,[(X,Y)=kb←query(end-corridor((X₂,Y₂),O₂)),
                   goto-landmark(X,Y),
                   goto-landmark(X₂,Y₂)])],
           [(L,2)<(F,1),(F,3)<(L,3)] ) )

( name:   blocking2b,
  roles, sit, res, ach: as in blocking2a
  body:   ([ (F, [gripper-down,open-gripper,gripper-up]),
             (L, [gripper-down,close-gripper,gripper-up,
                   store-box(B₁,S)])],
           [(F,1)<(L,2),(L,2)<(F,2),(F,2)<(L,3),(F,3)<(L,4)]))
```

Fig. 4.20. Joint plans that resolve the higher-level goal conflict

```
( name:   blocking3
  roles:  [F:FREE L:LOCKED]
  sit:    pos(F)=(X₁,Y₁,O₁) ∧ pos(L)=(X₂,Y₂,O₂) ∧
          ¬carries-box(F,_) ∧ carries-box(L,b₂) ∧
          between((X₂,Y₂),(X₃,Y₃),(X₁,Y₁,O₁))
  res:    blocked(L,leave-corridor) ∧
          blocked(F,goto-landmark(X₃,Y₃))
  ach:    achieves(F,load-ramp(R,B₂)) ∧
          achieves(L,load-ramp(R,B₁))
  body:   ([(F, [gripper-down,close-gripper,gripper-up,
                   load-ramp(R,B₂)]),
             (L, [gripper-down,open-gripper,gripper-up,
                   load-ramp(R,B₁)])],
           [(L,1)<(F,2),(F,2)<(L,2),(L,2)<(F,3),(F,3)<(L,4),(L,3)<(F,4)]))
```

Fig. 4.21. Joint plan involving the exchange of goals among agents

According to Definitions 3.6.5 and 3.6.8, each entry in the joint plan library has a name, a role declaration, an applicability condition, a *resolves*–condition, an *achieves*–condition, and a body, the latter consisting of a description of the plan for each role and of a set of precedence constraints between the different plans.

For example, consider Figure (4.12.b), assuming the following definition of the MAPP (see Definition 3.6.2) from the perspective of agent a_1:

M_1=({a_1,a_2},S_1,((leave-corridor),(goto-landmark((2,4)))))

with

S_1={pos(a_1)=(2,4,e) $\wedge$ pos(a_2)=(3,4,w) $\wedge$
        carries-box(a_1,b_1) $\wedge$ carries-box(a_2,b_2), ...}.

The plan blocking1 shown in Figure 4.19 is applicable to M_1 according to Definition 3.6.6: the situation description sit matches S_1, the leaf nodes of the goal paths of a_1 and a_2 match against the *resolves*–condition res in blocking1, and the *achieves*–conditions of blocking1 are dominated by the goal paths specified in M_1. The plan body of blocking1 implements the following conflict resolution procedure: the agent with the role FREE (i.e., a_2) clears the way for agent LOCKED, waits until LOCKED has left, and then continues in its plan. Note that neither of the plans shown in the Figures 4.20 or 4.21 are applicable to M_1.

As another example, consider Figure (4.12.c) and assume that the underlying multiagent planning problem is

M_2=({a_1,a_2},S_1,((park,leave-corridor),
        (unload-ramp(r,b_2)),store-box(b_2,s),goto-landmark((2,4)))))

with S_2={pos(a_1)=(2,4,e) $\wedge$ pos(a_2)=(3,4,w) $\wedge$ carries-box(a_2,b_2) $\wedge$...}.
Alike blocking1, the plans blocking2a and blocking2b are applicable to M_2. Whereas M_1 did not convey any information about the higher-level goals of the agents, the goal paths in M_2 reveal that agent LOCKED wants to leave the corridor in order to go parking, while FREE is looking for a place to store a box. According to plan blocking2a, the blocked agent moves one step back in order to allow agent FREE to achieve its goal of storing box b_1. Plan blocking2b suggests that agent LOCKED should adopt the goal of storing box b_1 from agent FREE; for this purpose, the latter agent has to hand over the box to the former. Which of both plans to choose is decided by plan evaluation (see Section 3.6.4).

As a third example, plan blocking3 can be applied to the situation shown in Figure (4.12.d) given by the MAPP

M_3=({a_1,a_2},S_3,((load-ramp(r,b_1)),store-box(b_1,r),leave-corridor),
        (load-ramp(r,b_2),search-box(b_2,s),goto-landmark((2,4)))));

according to this plan, both agents switch their top-level goals.

The last argument of a joint plan is a set of precedence constraints. They describe temporal dependence between the courses of action described by the individual role plans. For example, the set of precedence constraints for blocking1 is [(F,1)<(L,1),(L,1)<(F,2)]), saying that the first plan step of the agent with role variable F (i.e., agent a_2 in Figure (4.12.b)) must be carried out before the first plan step of the agent taking role L, which must be done before the second plan step of F.

Negotiation. The agreement on a joint plan to be executed is achieved according to Section 3.6.2. Using an election protocol, the two agents agree on a leader, who selects the protocol, assigns roles to the participants, and generates the negotiation set. This is done by applying the joint plan selection function (see page 111) to the plan library, returning a set of candidate plans. Then, joint plan negotiation is initiated by the leader by announcing the negotiation set and generating the first proposal using the joint plan negotiation protocol described in Section 3.6.2. For example, consider the conflict shown in Figure (4.12.c). Assume that the plan selection function determines the plans {blocking1, blocking2a, blocking2b} as candidates for conflict resolution. The role FREE is assigned to agent a_2, and LOCKED is assigned to a_1. Assume that a_1 has been elected leader. Then, according to the strategy of the leader in joint plan negotiations (see page 104), a_1 proposes the plan that has the maximal local utility for a_1. Assume now the following cost and worth function for a_1 and a_2, respectively:

cost functions:
```
ĉ₀(manipulator operations)=ĉ₀(kb-query)=ĉ₀(turning operations)=1
ĉ₀(send)=1,ĉ₀(wait)=0
ĉ₀(goto-landmark(Dest))=ĉ₀(gotoarea(Dest))=5 · dist(Curr-pos,Dest)
ĉ₀(store-box(B,S))=2 · #squares of S /* search place to store box */
```
worth functions:
```
w₀((leave-corridor))= 10
w₀((goto-landmark(X,Y)))=10
w₀((unload-ramp(R,B))=20)
```

$\hat{c}_0$ describes the underlying cost function for primitive plan steps (see Definition 3.5.3). Function w_0 (see Definition 3.6.12) returns the worth of a goal path for an agent. Based on these functions, the utility of a joint plan for an agent can be computed according to Definition 3.6.9 as $u(a_i, P) = w(a_i, P) - c(a_i, P)$. In Section 4.1, we made the assumption that both agents use the same utility functions; thus, we have $u(a_i, P) = w(P) - c(P)$.

In Section 3.6.3, two pairs of cost and worth functions were defined, resulting in two different utility functions. The functions c_1 and w_1 (see Definitions 3.6.10 and 3.6.13 on pages 115 and 116, respectively) only take into account the local cost and worth arising to the agent by the plan. The functions c_2 and w_2 (see Definitions 3.6.11 and 3.6.14 on pages 115 and 116, respectively) compute the global cost and worth of a joint plan, respectively. In our example, we set the weights ϵ_1 and ϵ_2 in the definition of the worth functions w_1 and w_2 to $\epsilon_1 = \epsilon_2 = 1$, i.e., the *resolves*–part and the *achieves*–part of the plan are weighed equally high. With $u_1(a, P) = w_1(a, P) - c_1(a, P)$ and $u_2(a, P) = w_2(P) - c_2(P)$, for the above example we have:

```
u₁(a₁,blocking1)=20-16=4,   u₁(a₁,blocking2a)=-2,   u₁(a₁,blocking2b)=0
u₁(a₂,blocking1)=20-26=-6,  u₂(a₁,blocking2a)=22,   u₁(a₂,blocking2b)=24
u₂(a₍₁,₂₎,blocking1)=4-6=-2,u₂(a₍₁,₂₎,blocking2a)=20,u₂(a₍₁,₂₎,blocking2b)=24
```

According to the joint plan negotiation protocol presented in Section 3.6.2, and assuming utility function u_1 for both agents, a_1 proposes the plans in the order blocking1, blocking2b, blocking2a; the preference ordering for a_2 is blocking2b, blocking2a, blocking1. Thus, a_1 first proposes blocking1, which is rejected by a_2; a_2 counterproposes blocking2b, which is accepted by a_1. If both agents use utility function u_2, plan blocking2b is selected first by both agents without further bargaining.

The negotiation process terminates and returns a solution if the initial negotiation set was not empty (see Theorem 3.6.1). The result of the negotiation is a joint plan, in this example blocking2b, which is then executed.

Plan transformation and execution. Committing itself to a joint plan has the following consequences for an INTERRAP agent: firstly, it computes the projection of its part from the joint plan. Secondly, it initiates the execution of the projected single-agent plan at the local planning layer. Finally, after the execution, it evaluates the state of conflict resolution. In the following, we describe the implementation of these steps in the loading dock.

Plan transformation:. In the two-agent case, a joint plan is transformed into two single-agent plans using the transformation rules defined in Section 3.6.3. For instance, the joint plan blocking2b for resolving the blocking conflict in Figure (4.12.c) is transformed into the following two plans for agent a_1 and a_2, respectively:

```
a₁:  [ gripper-down, wait(a₂),
       close-gripper, send(a₂),
       wait(a₂), gripper-up,
       wait(a₂), store-box(b₁, s) ]

a₂:  [ gripper-down, send(a₁),
       wait(a₁), open-gripper
       send(a₁), gripper-up,
       send(a₁) ]
```

The precedence constraints of the plan are implemented by the synchronizating commands wait and send (see page 112ff). The plan structures obtained by the transformation process are regular single-agent plans that are posted to the local planning layer, where they are interpreted.

Commitment to joint plans. Plans that are posted down to the local planning layer by the cooperative planning layer are scheduled within the current local plans. This is not trivial, as the effects of the plans have to be taken into account. For example, it is not always satisfying to replace the (leaf node) plan step during the execution of which the conflict occurred by the set of plan steps returned by the cooperative planning layer. The reason for this is that additional higher-level goals can become satisfied or obsolete as a result of the cooperative planning process, and that the agent's local plan stacks must be modified accordingly. For example, while plan blocking1 achieves exactly the navigation goals leave-corridor and goto-landmark of the conflicting agents in Figure (4.12.c), in plan blocking2b agent a_1 adopts the goal of storing the box from agent a_2; thus, while the effect of

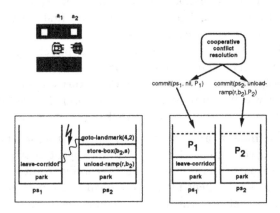

Fig. 4.22. Modification of local plan stacks by cooperation commitments

executing plan `blocking2b` for agent a_1 is that it can keep on achieving its goal to go parking (hopefully undisturbed) after doing some additional work, the effect for a_2 is that its whole top-level goal `load-ramp` is satisfied by the plan, and that it can delete the top-level plan from the plan stack and replace it by the conflict resolution plan.

In general, integrating the conflict resolution plan into the existing plan would require to analyze what goals are achieved by the joint plans. This, however, would require a full-fledged planning mechanism working on a state-based representation of goals. In the current implementation of the loading dock, no such mechanism is available. Therefore, in order to provide the required information, joint plans have a slot representing the expected post-conditions of their execution. Given the goal path $\{p_1, \ldots, p_n\}$ of an agent, which is sent to the cooperative planning layer as a description of the agent's mental model, the *achieves*–condition is of the form $p_i \in \{p_1, \ldots, p_n\}$, denoting the topmost plan step in the goal path that is satisfied by the plan.

The conflict resolution plan is posted down to the local planning layer as a message `commit(id, `p_i`, P)`, where

- id identifies the plan stack in the execution of which the conflict was recognized;
- $p_i \in \{p_1, \ldots, p_n\} \cup \{nil\}$ denotes the *achieves*–condition of the plan;
- P is the conflict resolution plan.

The operational semantics of a message `commit(id, `p_i`, P)` is the following:

- The plan stack denoted by id is cleared up to the element p_i including that element. If p_i = nil, no element is removed from the plan stack.
- Plan P is pushed on the plan stack denoted by id.

Figure 4.22 shows the plan stacks of both agents a_1 and a_2 before and after the integration of the returned projections of plan blocking2b. The plan stack ps_1 of a_1 consists of two goals: the goal park and the current subgoal leave-corridor. The postcondition of plan blocking2b for agent a_1 is nil, i.e., the execution of the joint plan does not directly achieve one of a_1's goals, but removes the conflict with the goal of a_2 so that the goal can then be achieved by continuing plan execution at the local planning layer. Thus, a_1's plan stack is modified by pushing plan P_1 on top of the stack ps_1 without deleting any element from it. The *achieves*-condition of blocking2b for a_2 is load-ramp(r,b_1). Thus, all elements on a_1's plan stack, including load-ramp(r,b_1) itself, can be deleted before pushing plan P_2. Thus, after the conflict resolution plan is executed, agent a_2 will continue by parking, leave the corridor and allowe a_1 to continue with its plan.

Handling different outcomes of cooperative conflict resolution. After the execution of the projection of a joint plan, the local planning layer returns an acknowledgment, on the basis of which the cooperative conflict resolution either succeeds or fails at the cooperative planning layer. In the former case, the success is reported to the layer that activated the cooperative conflict resolution, i.e., to the local planning layer, and the agent continues with its local tasks. In the latter case, the cooperative planning layer will search for alternatives; if none are found, it reports a failure to the local planning layer, where a solution must be found. In the current implementation, the local planning layer passes the failure to the behavior-based layer, where an emergency PoB is activated: the agent makes n random moves, where n is chosen depending on the size of the scenario, and then tries to resume its activites.

4.4 The FORKS Simulation System

The loading dock application was implemented in the FORKS simulation system. FORKS consists of four parts: the agent society, the simulation world, the graphical user interface (GUI), and an agent toolset. Figure 4.23 shows the top-level window of the graphical user interface of FORKS. The GUI allows the human user to interact with the simulation system, to adjust parameters of the simulation, to create and to destroy agents, to trace their execution, to instruct them interactively, and to observe their behavior in the simulation window. In the following, a few important aspects of the simulation system are briefly discussed.

Creating agents. Different types of agents can be created using the menu displayed in Figure 4.24. The user can select a host on which to run the agent's control program, assign a physical robot device to the agent, give it different basic capabilities, like curiosity, the ability to communicate, and some *a priori* knowledge about the world. Agents can be created that differ both in their local task planning abilities and in their conflict handling

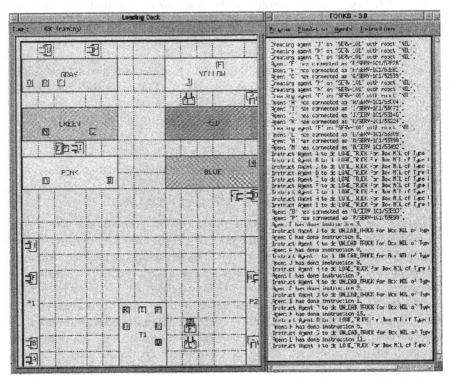

Fig. 4.23. The FORKS simulation system

behavior. Local task planning abilities range from random walk to planning using the plan library approach described in Section 3.5 (cooperative planning of local tasks is not possible in the current version, see Section 6.2 for a discussion). Conflict resolution strategies range from random moves to cooperative conflict resolution in the cooperative planning layer. In Section 5.4, we report on experiments with different agent types.

Tracing agent execution. The activity at the individual control layers of an INTERRAP agent can be observed by using the INTERRAP tracer, which is part of the agent toolset. It allows to dynamically trace and untrace individual agents. Figure 4.25 shows a snapshot of the simulation system with two agents *A* and *B*. The tracer window of an agent has four subwindows: (i) the world interface window that displays the primitive actions executed by the agent, (ii) the window for the behavior-based layer, providing information about active PoBs and conflicts recognized at that layer, (iii) the window for the local planning layer that documents the agent's goals and the process of plan

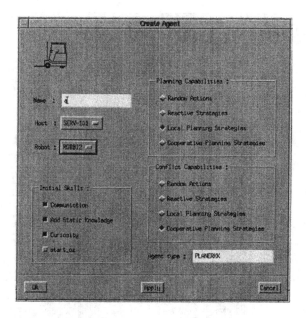

Fig. 4.24. Menu for creating agents

expansion, and (iv) the window of the cooperative planning layer, showing the state of protocol execution[9].

KHEPERA miniature robots. We have extended the FORKS system to be able to control physical KHEPERA robots. These robots (see Figure 4.26 and [MFI93]) are miniature robots that are equipped with a simple sensory system, two step motors, each 256KB RAM and ROM on-board, and a gripper module. The available sensors include eight infrared sensors that allow to detect obstacles, three (custom-made) floor infra-red sensors that the robot can use to follow a white line on the floor, and to orient itself that way. The gripper is equipped with a light barrier that enables the robot to test the success of gripper action, and it allows to measure the electrical resistance of gripped objects. This can be used to distinguish between different types of boxes. The motors can be controlled independent of each other; this enables a flexible movement of the robot such as driving curves forward and backward, and turning on the spot.

The layered architecture of INTERRAP has made the extension of the FORKS simulation system to physical robots easy in that no changes were required neither at the local nor at the cooperative planning layer. Only minimal changes had to be made at the behavior-based layer, in that a part of the reactor PoBs were reimplemented as subsymbolic, neural-net–like algo-

---

[9] For historical reasons, the control layers in the tracer are still labeled with the names they had in earlier versions of INTERRAP, see e.g., [MP94].

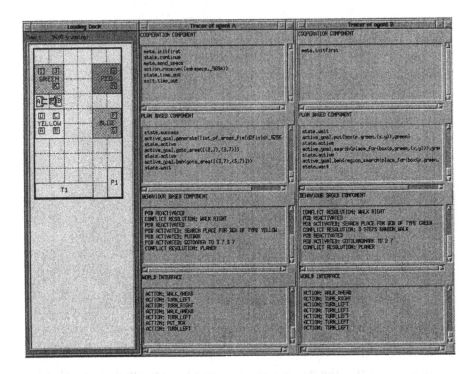

Fig. 4.25. The INTERRAP tracer

Fig. 4.26. KHEPERA miniature robots

rithms on the robot, allowing fast reaction to unforeseen events. However, these adaptations to the simulation system were achieved within less than a man-month.

Technical details. In the current version of FORKS, each INTERRAP agent consists of one to three parallel UNIX processes:

1. Each agent has a behavior-based layer and a world interface that are implemented using the rule-based language MAGSY [Fis93b] [Fis93a].
2. A PROLOG process implements the local planing layer and the cooperative planning layer of an agent. These layers are optional.
3. A physical robot additionally has a control process written in C, which is downloaded on the robot, and which communicates with the MAGSY control process. The robot control program implements subsymbolic PoBs that are directly based on sensory input, as well as control procedures for the navigation and manipulation actions described in Section 4.2.2.

Agents communicate via the TCP/IP protocol, which is run between the MAGSY processes. A wireless communication module for the robot agents is available, but has not been integrated so far. Recently, the implementation of the agent specification language ALADIN (A Language for Designing INTERRAP Agents) based on the INTERRAP architecture and written in the Oz language [HSW93] became available (see [RMP95], [Ros96]). ALADIN provides a domain-independent tool supporting the design of INTERRAP agents in a uniform high-level language framework. The FORKS system is currently being reimplemented in ALADIN.

4.5 Bottom Line

The main concepts of the INTERRAP architecture were developed and evaluated in the simulation of an automated loading dock inhabited by forklift robots. This application allows us to study the reconciliation of reactivity and deliberation, and different kinds of mechanisms for conflict handling supported by INTERRAP. Two general types of conflict resolution mechanisms were presented: behavior-based mechanisms do not require communication among agents, and allow to resolve a variety of conflict situations by local decision-making. Mechanisms based on cooperative planning allow agents to coordinate their actions by exchanging goal information, and by agreeing on joint plans to resolve conflicts. The resulting simulation system FORKS is a testbed that allows the user to experiment with a variety of (simulated or physical) robot agents. It was presented at various industrial fairs (e.g., CeBIT'94 and '95) and conference exhibitions (e.g., KI-94, ICMAS-95).

5. Evaluation

5.1 Introduction

The INTERRAP architecture offers the designer of a multiagent system a wide variety of alternatives for modeling agents in a multiagent system, using different reactive or deliberative planning mechanisms, and employing different local or communicative mechanisms to deal with interactions with other agents. The question that we investigate in this chapter concerns the effects that different design decisions at the level of the individual agent have on the multiagent system as a whole.

The interaction behavior of INTERRAP agents—and of agents in general— can be classified according to two dimensions shown in Figure 5.1. The first dimension is that of coordination, which can be done either by communication with other agents or by local, non-communicative mechanisms[1]. The second dimension is that of decision-making: are reactive decision-making mechanisms used, or are the decisions of an agent based on a considerable amount of deliberation? Figure 5.1 illustrates different instances and techniques resulting from the varying combinations of both dimensions, which themselves describe continuous transitions rather than being two extremes.

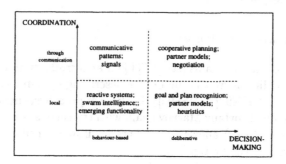

Fig. 5.1. Dimensions of interactive agent behavior

---

[1] See [Ros85] [RZ94] for game-theoretic interpretations of this classification.

Our analysis focuses on the dimension of coordination. Using the loading-dock application as an example, our goal is to achieve some insights into the effects of local and communicative forms of coordination on the performance of the individual agent in a group of agents. We use two analytic techniques to achieve this goal: Firstly, we present a model based on finite Markov chains that allows us to predict the performance of agents that use probabilistic decision functions (see Definition 4.3.1 in Section 4.3.4) to perform navigation tasks. Secondly, we evaluate the performance of different types of INTERRAP agents by presenting experimental results obtained by the FORKS computer simulation; this will allow us to also consider agents whose behavior is based on communication and cooperative planning, which are not accessible to the analysis based on Markov chain theory.

The chapter is structured as follows: Section 5.2 analyzes the interactive behavior of a specific type of INTERRAP agents using the tools of Markov chain theory. In Section 5.3, the general model presented in Section 5.2 is applied to the loading dock application. Section 5.4 presents and evaluates a set of experimental results. The scope and limitations of the models are discussed in Section 5.5.

5.2 A Model for Behaviour-Based Decision-Making

The main idea we suggest in the following is to use the model of finite Markov chains to approximately predict the collective behavior of interacting behavior-based robots the individual decisions of which are based on prob-abilistic decision functions (see Section 4.3.4). In particular, these decision processes have the Markov property (see below). This holds true because taking only the current state of the world into account for making decisions is suitable to achieve reactive and situated behavior, which is the main purpose of the behavior-based layer.

5.2.1 Finite Markov chains

Recently, finite Markov chains (see [KS60] for an introduction) have been used in the planning literature as a model for describing dynamic uncertain environments (see [DKKN93] [THSS95]). In these approaches, encoding planning problems into an absorbing Markov chain with transient states and absorbing states (the goal states) allows researchers to draw quantitative conclusions on the behavior of the system.

The way we are using Markov chains is not only to cover the uncertain *outcome* of actions (although this will play a role in coping with the presence of multiple agents) but also to quantify properties of the class of probabilistic decision functions defined in Section 4.3.4, which describe the *choice* among actions as a stochastic process.

A *finite Markov process* is a finite stochastic process describing probability transitions trough a set of states whose conditional probability distribution satisfies the *Markov property*: the transition probability p_{ij} from state s_i to state s_j depends only on s_i, but not on any previous state. A *finite Markov chain* is a finite Markov process whose transition probability p_{ij} is independent of how often state s_i was reached in the past. A Markov chain whose probability does not change over time is called *stationary*. An *absorbing chain* is defined by two disjoint sets of states: transient states and absorbing states. Absorbing states are those states that are never left once entered.

The transition matrix M_{ij} of a finite absorbing Markov chain M has the form

$$\begin{pmatrix} Q & R \\ 0 & I \end{pmatrix} \tag{5.1}$$

where Q are the transitions among transient states, R are transitions from transient to absorbing states, 0 contains only zeros, and I is the identity. The following proposition recalls results of Markov chain theory [KS60]:

Proposition 5.2.1. *Let* $M = \begin{pmatrix} Q & R \\ 0 & I \end{pmatrix}$ *be a finite stationary Markov chain. Then:*

- $N \overset{\text{def}}{=} (I - Q)^{-1} = \sum_{t=0}^{\infty} Q^t$ *always exists; given an initial state* s_i, *for any transient state* s_j, N_{ij} *is the average number of times* s_j *is entered before an absorbing state is reached.* N *is called the fundamental matrix to* M.
- $V \overset{\text{def}}{=} N \cdot (2 \cdot N^d - I) - N^2$ *describes the variance for* N, *i.e.,* V_{ij} *is the variance of the total number of times the process is expected to be in a transient state* s_j *if the chain started in* s_i. N^2 *is computed by squaring each entry of* N; N^d *results from* N *by setting each off-diagonal entry of* N *to zero.*
- $(N \cdot R)_{ij}$ *computes the probability of reaching the absorbing state* s_j *if the initial state was* s_i.

5.2.2 A probabilistic model for agent interaction

In the following, we procede in three steps: firstly, we provide the necessary formal framework by adopting the general Markov model to our settings; secondly, we show how this setting can be used to predict the behavior of an indidivual decision-making and acting agent; thirdly, we discuss two ways of representing the effects of agent interaction into the model.

Formal framework. We define a finite absorbing Markov chain as a tuple $M \overset{\text{def}}{=} (\mathcal{S}, \mathcal{ACT}, P)$ where $\mathcal{S} = \mathcal{S}_t \cup \mathcal{S}_a, \mathcal{S}_t \cap \mathcal{S}_a = \emptyset$ is a finite set of world states, $\mathcal{S}_t$ represents transient world states and $\mathcal{S}_a$ represent absorbing world states given by the set $\mathcal{G}$ of goal states (throughout the paper, we shall mostly deal with the case where $\mathcal{S}_a$ is singleton; this, however, is not a general restriction); $\mathcal{ACT} = \{a_1, \ldots, a_k\}$ is a finite set of actions; $P : \mathcal{S} \times \mathcal{S} \mapsto [0; 1]$

is a probabilistic transition function of the environment: $P(s_i, s_j)$ denotes the probability of getting from state s_i to state s_j. In order to express that P depends on the set of goal states S_a, we write P^{S_a} or simply P^g if $S_a = \{g\}$.

In our application, the function $P \equiv P(f, e)$ is obtained by the composition of two probability distributions: the probability distribution f over possible action selections in a situation, and the distribution e over the possible outcomes of executing an action. The function

$$ f : \mathcal{S} \times \mathcal{ACT} \times \mathcal{S}_a \mapsto [0; 1] $$

is the function underlying the actual decision function $\mathcal{F}$ of an agent (see Definition 4.3.1). For $s \in \mathcal{S}$, $a \in \mathcal{ACT}$, $g \in \mathcal{S}_a$, $f(s, a, g)$ denotes the probability that a is selected for execution by an agent in state s given the agent's current goal is g. The function

$$ e : \mathcal{S} \times \mathcal{ACT} \times \mathcal{S} \mapsto [0; 1] $$

describes a probability distribution over the possible outcomes of actions: for $s_1, s_2 \in \mathcal{S}$, $a \in \mathcal{ACT}$, $e(s_1, a, s_2)$ is the probability of reaching s_2 by executing a given the initial state s_1.

Behavior-based decision-making. Given an agent that makes its decisions using a PDF $\mathcal{F}^f$, the above framework can be used as a model to predict the agent's behavior. For this purpose, we define the transition function P:

Definition 5.2.1. *Given the initial world state s_0 and a goal state g, the transition function $P^g : \mathcal{S} \times \mathcal{S} \mapsto [0; 1]$ is given by the transition matrix P_{ij}^g as the transition probability from state s_i to state s_j with*

$$ P_{ij}^g = \sum_{\substack{a_k \in \mathcal{ACT} \mid a_k \text{ has} \\ \text{been selected in } s_i}} e(s_i, a_k, s_j) = \sum_{a_k \in \mathcal{ACT}} f(s_i, a_k, g) \cdot e(s_i, a_k, s_j). $$

By dividing the set of states into two subsets $\mathcal{S}_t$ and $\mathcal{S}_a$, where $\mathcal{S}_a = \mathcal{G}$ and $\mathcal{S}_t = \mathcal{S} - \mathcal{G}$, P_{ij}^g can be written in the canonical form according to Equation 5.1. By computing the fundamental matrix N and its variance matrix V in accordance with Proposition 5.2.1, we can determine e.g., the expected number of actions n_i the agent will perform before it reaches a state $g \in \mathcal{G}$ starting from state s_i, which is $n_i = \sum_{j=1}^{n-1} N_{ij}$ if S_a is singleton (see the example in Section 5.3).

Interaction by local behavior. So far, we can draw quantitative conclusions about the performance of a problem-solving agent which is alone in the world. Next, we shall extend our framework to deal with the presence of multiple agents. In the following, we shall discuss two ways of integrating the presence of other agents into our framework: a probabilistic model and an exact model, the latter taking the current situation into account.

A naive model. For the probabilistic model, we make use of the observation that at the level of local behavior-based decision-making, the main effect of the presence of others is that it increases the likelihood of failure of domain actions. A probabilistic extension of the above model is proposed that allows us to predict the approximate performance of an agent given the size of its environment and the number of other agents being around.

Given a Markov chain model $M = (S, \mathcal{ACT}, P(f, e))$ as defined above, the main idea is to incorporate knowledge about the effects of the presence of other agents into the transition function P. For the scope of this paper, we restrict ourselves to a basic type of interaction which occurs in the domain we are looking at: the probability of an action to fail increases linearly with the number of other agents being present: for k agents, the probability for an action a by an individual agent to fail in state s is $Pr(fails(a, s)) = \frac{k-1}{c}$ for a constant $c \geq k - 1$. In order to cover the notion of execution failure which occurs if an action is carried out whose precondition is not satisfied, we make the following assumption[2]:

Definition 5.2.2 (Execution failure). *Let $s \in S$ be a state of the world, $a \in \mathcal{ACT}$ be an action. Let $prec(a)$, $exec(a, s)$ be formulae denoting the pre-conditions of a, and the execution of a in state s, respectively; let $\mathrm{eff}(a, s)$ the state resulting from executing a in state s, and let $holds(p, s)$ be a meta predicate evaluating to true if predicate p holds in state s. Then, $\neg holds(prec(a), s) \land exec(a, s) \Rightarrow \mathrm{eff}(a, s) = s$.*

The implication for the transition function P is that the failure of the execution of an action will result in an unchanged state, i.e., a transition $s \to s$. This is covered by the following definition:

Definition 5.2.3. *Given the local $n \times n$ transition matrix $P = P^1$ for one agent. Then the k-agent transition matrix P^k, $k \geq 2$, is defined by*

$$
P_{ij}^k = \begin{cases} \frac{k-1}{c} \cdot \sum_{\substack{m=1 \\ m \neq i}}^{n} P_{im} & \text{if } i = j \\ \left(1 - \frac{k-1}{c}\right) \cdot P_{ij} & \text{otherwise} \end{cases}
$$

Now, we are able to define the naive model as a Markov chain.

Definition 5.2.4 (Naive model). *Let $M = (S, \mathcal{ACT}, P)$ be a finite absorbing Markov chain for an individual agent; let k be the number of agents. Then $M^k = (S, \mathcal{ACT}, P^k)$ is the Markov chain denoting the naive model of interaction.*

Theorem 5.2.1. *Let M be a finite absorbing Markov chain. Then M^k is a finite absorbing Markov chain.*

In Section 5.3, we shall give an example of how the naive model can be used to derive certain quantities of the process.

---

[2] This restriction is made for simplicity. It could be weakened by assuming a probability distribution over different possible states.

An exact model. The main problem of the approach based on approximation is that it does not allow good predictions of the performance of agents given specific initial situations, such as conflicts (see Figure 5.2). Therefore, we provide an alternative approach, which models the system consisting of several agents by one Markov chain. The Markov transition matrix for the system can be computed from the transition matrices of the individual agents. It is important to note that in constructing the exact model we raise our view from the perspective of the individual agent to a global perspective of the system. We start from a multiagent system with k agents, in the following denoted by indices $1 \leq i \leq k$, where the local behavior of each agent is described by a finite absorbing Markov chain.

Definition 5.2.5. *Let $A = \{1,\ldots,k\}$ be a set of agents. For each $i \in A$, let $M^i = (S^i, ACT^i, P^i)$ be the Markov chain describing the local behavior of i. Let D be a set of first-order formulae denoting a domain theory. Then, $M = (S, ACT, P)$ is the Markov chain denoting the exact model, with $S = S^1 \times \ldots \times S^k$, $ACT = ACT^1 \times \ldots \times ACT^k$, $P = \gamma(D, P^1, \ldots, P^n)$.*

States are defined by the Cartesian product of the states of the individual agents; states change due to simultaneous actions of the agents. The transition probabilities are computed by the probabilities of the individual agents using a function γ. A naive way to derive the probability of a transition among two states $s, s' \in S$, $s = (s_1 \ldots s_k)$ and $s' = (s'_1 \ldots s'_k)$ is to define $P(s, s') = \prod_{i=1}^{k} P_i(s_i, s'_i)$. However, we have to exclude inconsistent state transitions using a domain theory D representing a set of domain constraints. D partitions the set S of states into two disjoint subsets, the admissible and the inconsistent state transitions:

Definition 5.2.6. *Let $M = (S, ACT, P)$ be a finite absorbing Markov chain for k agents. Let D be a domain theory. Let $s, s' \in S$, $s = (s_1 \ldots s_k), s' = (s'_1 \ldots s'_k)$. Then the set of admissible state transitions AS is defined as $\{s \to s' \in S | P^i(s_i \to s'_i) > 0 \wedge \text{consistent}(D, s \to s')\}$. The set of inconsistent state transitions is $IS = S - AS$.*

I.e., for a state transition to be admissible we require both the feasibility of the local state transitions and consistency of the global state transition with the domain constraints. In case the global transition achieved by the set of simultaneous local transitions is inconsistent, the system as a whole is still required to end up in a well-defined state. Therefore, we define a set CS of *compromise state transitions* whose purpose is to replace inconsistent state transitions. CS characterizes the set of alternative states which can be taken by the k-agent system whenever the simultaneous local decisions of the agents would lead to a globally inconsistent state $\hat{s} \in IS$. CS is required to be consistent with D; however, for $s \to s' \in CS$, we do not require that $P^i(s_i, s'_i) > 0$. I.e., the result of a compromise state transition for an individual agent can be one the agent herself would never have selected if it was the only agent in the world.

Definition 5.2.7. *Algorithm for computing $P = \gamma(P^1, \ldots, P^k)$:*

1. *Compute the Cartesian product of all possible state transitions: $S' := S^1 \times \ldots \times S^k$.*
2. *$\forall s, s' \in S', s = (s_1, \ldots, s_k), s' = (s_1', \ldots, s_k') : p'(s, s') := \prod_{i=1}^{k} P^i(s_i, s_i')$.*
3. *Compute AS and IS from S'. Set $CS := \emptyset$.*
4. *$\forall (s, s') \in AS: P_{AS}(s, s') := p'(s, s')$.*
5. *$\forall (s, s') \in IS: CS := CS \cup \tau(s, s')$, where $\tau(s, s')$ is the set of all consistent subsequent states of s when trying to reach state s'.*
6. *For all $(s, s') \in IS$ and for all $\hat{s} \in \tau(s, s') : P_{CS}(s, \hat{s}) := p'(s, \hat{s}) + \frac{p'(s,s')}{|\tau(s,s')|}$.*
7. *For all $(s, s') \in IS$ set $P_{IS}(s, s') := 0$.*
8. *$P = P_{AS} \cup P_{CS} \cup P_{IS}$*

Theorem 5.2.2. *Let $\mathcal{A}$, M^i, $\mathcal{D}$, and $M = (\mathcal{S}, \mathcal{ACT}, P)$ be as in Definitions 5.2.5 and 5.2.7. Let M^i describe a finite absorbing Markov chain for each $i \in \mathcal{A}$. Then M describes a finite absorbing Markov chain.*

Theorem 5.2.2 guarantees us that, having computed P, the model $M = (\mathcal{S}, \mathcal{ACT}, P)$ can be used in the standard way by applying the tools of Markov chains to derive quantitative properties of the multiagent system. For an example, we refer to Section 5.3.

Whereas the exact model allows us to make detailed predictions on the future behavior of the system, this analysis is not for free. Whereas space and time complexity of the naive model are polynomial (space complexity in $\mathcal{O}(n^2)$, time complexity in $\mathcal{O}(n^3)$), the space requirement for the exact model for n agents with m local states is in $\mathcal{O}(n^{2 \cdot m})$. The complexity of computing the fundamental matrix is in $\mathcal{O}(n^{3 \cdot m})$ and thus exponential in the number of agents.

However, in many domains (in particular in the one investigated here), many interesting situations (such as the conflicts described in Section 4.3) can be described by rather simple settings and do not involve a prohibitive number of states. In these cases, as we shall see in the following, the exact model is tractable and useful to analyze global effects of local decision algorithms in specific situations. Thus, it can help the system designer to avoid certain dangerous pitfalls in the local modeling of agents.

5.3 Behaviour-Based Interaction in FORKS

In this section, we shall apply the general model defined in Section 5.2 to the behavior-based navigation techniques presented in Section 4.3.4. We instantiate both the naive and the exact model and analyze the performance of agents equipped with behavior-based navigation techniques in a multiagent environment.

5.3.1 The model

First, we shall map probabilistic decision functions for individual agents into a Markov chain by defining states, actions, and state transitions.

States:. The part of a forklift's state that is relevant for our model is a 3-tuple (x, y, o), where x and y denote its x- and y-coordinates, respectively, and $o \in \{s, n, e, w\}$ its direction. For the navigation task considered in this section, we can even further restrict the state space for the Markovian model to the set of possible tuples (x, y).

Actions:. The possible actions that have been defined in Section 4.3.4 is the *moveto* action which makes the forklift move to a direction specified as an argument to the action. The operational semantics of *moveto* is as follows:

```
Name:          moveto(Dir)
Effects:       S = ((X,Y,O) ↦ ((X',Y',O') with
               O' = Dir      /* the forklift can only move forward */
               (X', Y') = switch (Dir)      /* case distinction */
               {
                   case 'n': (X, Y+1);
                   case 'e': (X+1, Y);
                   case 's': (X, Y-1);
                   case 'w': (X-1, Y)
               }
Precond:       type((X', Y')) = ground ∧
               status((X', Y')) = nil
```

Here, `type` and `status` are the attributes describing a square as defined in Section 4.2. Thus, the action `moveto(D)` is applicable if the forklift believes that the field denoted by the direction `D` is free. As the belief of a forklift may not coincide with the real world, this definition describes the intended effects of the action rather than the actual effects at the time of execution.

Transitions:. Transitions among local states that correspond to movements of the forklifts through the scenario are achieved by the execution of actions. The incentive to act is given by goals. In our example, a goal is given by a location the robot is to reach. This goal is explicitly represented at the local planning layer and operationalized by calling the procedure PoB `goto-landmark-pdf` (see Figures 4.4 and 4.17 in Chapter 4). Thus, there is a natural mapping from goals to states.

The local transition function P^i of agent i is given by instantiations of the functions f^i and e^i: The function e^i that represents the outcome of an action is assumed to be deterministic. I.e., for the one-agent case we assume that actions bring about the expected effects. f^i defines the probabilities of the robot to move to some direction depending on the relative position of the goal relative to the robot's current position. The underlying decision function $\mathcal{F}^f$ is based on the notion of quadrants (see Figure 4.18 on page 160). Given the current position s and a goal g, the world is divided into four quadrants $q_1 - q_4$. q_1 is the quadrant where the goal field is located. q_4 is the quadrant which is directed away from the goal; q_2 and q_3 are orthogonal to the goal. Given a state s and a goal state g, $q_i(s,g)$ returns the state (square) which is directly reachable from s and which is in quadrant q_i relative to g. Thus, the square $q_i(s,g)$ is selected with a probability of $f(s,q_i,g)$. In the following, let q_i denote the action which will result in the state of the agent being $q_i(s,g)$. $f(s,q_i,g)$ can be equal to zero if the agent cannot move to this direction because there is an obstacle there. For this case, we include a dummy operator $q_0(s,g) = s$. Function P^i can now be defined as follows:

Definition 5.3.1. *Let $\mathcal{F}^f$ be a PDF; let $c_1, c_2 > 1$. Let $\mathcal{ACT}^i = \{q_j | 0 \leq j \leq 4\}$. For all s, g, let f be constrained by the following conditions:*

(1) $\sum_{i=0}^{4} f(s, q_i, g) = 1$
(2) $f(s, q_1, g) > 0 \wedge f(s, q_2, g) > 0 \Rightarrow f(s, q_1, g) = c_1 \cdot f(s, q_2, g)$
(3) $f(s, q_1, g) > 0 \wedge f(s, q_3, g) > 0 \Rightarrow f(s, q_1, g) = c_1 \cdot f(s, q_3, g)$
(4) $f(s, q_2, g) > 0 \wedge f(s, q_3, g) > 0 \Rightarrow f(s, q_2, g) = f(s, q_3, g)$
(5) $f(s, q_2, g) > 0 \wedge f(s, q_4, g) > 0 \Rightarrow f(s, q_2, g) = c_2 \cdot f(s, q_4, g)$
(6) $f(s, q_3, g) > 0 \wedge f(s, q_4, g) > 0 \Rightarrow f(s, q_3, g) = c_2 \cdot f(s, q_4, g)$
(7) $f(s, q_0, g) > 0 \Leftrightarrow \sum_{i=1}^{4} f(s, q_i, g) = 0$

Furthermore, let $e^i : S \times \mathcal{ACT} \times S \mapsto [0; 1]$, $e_i(s, q_i, q_i(s,g)) = 1$. Then, $P^i : S \times S \mapsto [0; 1]$ is defined by

$$P^{i,g}(s, s') = \sum_{q \in \mathcal{ACT}} f_i(s, q, g) \cdot e_i(s, q, s') = \sum_{q \in \mathcal{ACT}} f_i(s, q, g).$$

For $c_1 = c_2 = 1$, we obtain a random walk strategy, where the agent selects each direction with the same equality. For $c_1, c_2 > 1$, we obtain a weighted strategy that makes the agent move towards its goal with a higher probability.

Example:. Let us consider Example (5.2.a). Assume that agent a who is at

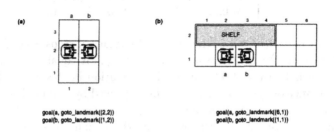

Fig. 5.2. Conflict situations in the loading dock

location $(1, 2)$ has the goal to move to location $(2, 2)$. Then, $q_1((1,2), (2,2)) = (2,2)$, $q_2((1,2), (2,2)) = (1,1)$, $q_3((1,2), (2,2)) = (1,3)$; q_4 is not possible in the example. By choosing $c_1 = 4$, $c_2 = 1.5$, we obtain the probabilities $p((1,2),(2,2)) = \frac{2}{3}$, $p((1,2),(1,3)) = p((1,2),(1,1)) = \frac{1}{6}$. Thus, the canonical forms of the local transition matrices P^a and P^b are as follows:

$$
P^a = \begin{array}{c|cccccc}
 & 11 & 12 & 13 & 21 & 23 & 22 \\
\hline
11 & 0 & \frac{1}{5} & 0 & \frac{4}{5} & 0 & 0 \\
12 & \frac{1}{6} & 0 & \frac{1}{6} & 0 & 0 & \frac{2}{3} \\
13 & 0 & \frac{1}{5} & 0 & 0 & \frac{4}{5} & 0 \\
21 & \frac{1}{5} & 0 & 0 & 0 & 0 & \frac{4}{5} \\
23 & 0 & 0 & \frac{1}{5} & 0 & 0 & \frac{4}{5} \\
22 & 0 & 0 & 0 & 0 & 0 & 1
\end{array}
$$

$$
P^b = \begin{array}{c|cccccc}
 & 11 & 13 & 21 & 23 & 22 & 12 \\
\hline
11 & 0 & 0 & \frac{1}{5} & 0 & 0 & \frac{4}{5} \\
13 & 0 & 0 & 0 & 0 & \frac{1}{5} & \frac{4}{5} \\
21 & \frac{4}{5} & 0 & 0 & \frac{1}{5} & 0 & 0 \\
23 & \frac{1}{6} & 0 & \frac{1}{6} & 0 & 0 & \frac{2}{3} \\
22 & 0 & \frac{4}{5} & 0 & \frac{4}{5} & 0 & 0 \\
12 & 0 & 0 & 0 & 0 & 0 & 1
\end{array}
$$

Based on the single-agent Markov chains, we shall now provide examples for the naive model and the exact model defined in Section 5.2.

5.3.2 Analysis of system behavior

The naive model. The transition matrix P^a for agent a in the example shown in Figure (5.2.a) can be used to derive quantities of the local problem-solving behavior of agent a. The fundamental matrix N^a for P^a (see Definition 5.2.1) has the form

$$
N^a = \begin{array}{c|ccccc|c}
 & 11 & 12 & 13 & 21 & 23 & total \\
\hline
11 & 1.24 & 0.26 & 0.05 & 0.99 & 0.04 & 2.59 \\
12 & 0.22 & 1.09 & 0.22 & 0.18 & 0.18 & 1.88 \\
13 & 0.05 & 0.26 & 1.24 & 0.04 & 0.99 & 2.60 \\
21 & 0.25 & 0.05 & 0.01 & 1.20 & 0.01 & 1.52 \\
23 & 0.01 & 0.05 & 0.25 & 0.01 & 1.20 & 1.52
\end{array}
$$

Thus, for $c_1 = 4, c_2 = 1.5$, the expected number of steps needed to reach the goal state $(2,2)$ when starting from state $(1,1)$ is equal to $\bar{n} = 2.59$, the variance is equal to $v = 1.08$. The optimal solution in this case is 2. Thus, the performance of agent behavior can be improved by making it more goal-directed, i.e., by selecting other values for c_1, c_2. For $c_1 - c_2 \to \infty$, the solution asymptotically approximates the optimum; the variance decreases. E.g., for $c_1 = 20, c_2 = 4.5$, we have $\bar{n} = 2.11$ and $v = 0.21$. However, as we shall see, if we introduce multiple agents which all behave according to this strategy, this is no longer the case.

Yet, first we provide an instantiation of the naive model defined above. For k agents and a number of grids of n, we compute the k-agent matrix P^k

according to Definition 5.2.3, setting $c = n - 1$ and $p(fails(a, s)) = \frac{k-1}{n-1}$. For Example (5.2.a) and $k = 5$, $c_1 = 4$, $c_2 = 1.5$, $P^{k,a}$ and $N^{k,a}$ are computed as:

$$P^{k,a} = \begin{pmatrix} 0.8 & 0.04 & 0 & 0.16 & 0 & 0 \\ 0.03 & 0.8 & 0.03 & 0 & 0 & 0.13 \\ 0 & 0.04 & 0.8 & 0 & 0.16 & 0 \\ 0.04 & 0 & 0 & 0.8 & 0 & 0.16 \\ 0 & 0 & 0.04 & 0 & 0.8 & 0.16 \end{pmatrix}$$

$$N^{k,a} = \begin{pmatrix} 6.18 & 1.28 & 0.23 & 4.95 & 0.18 & \mathbf{12.82} \\ 0.96 & 5.38 & 0.96 & 0.77 & 0.77 & \mathbf{8.84} \\ 0.23 & 1.28 & 6.18 & 0.18 & 4.95 & \mathbf{12.82} \\ 1.24 & 0.26 & 0.05 & 5.99 & 0.04 & \mathbf{7.58} \\ 0.05 & 0.26 & 1.24 & 0.04 & 5.99 & \mathbf{7.58} \end{pmatrix}$$

For getting from $(1,1)$ to $(2,2)$ in the above example, we have $\bar{n} = 12.82$; the variation $v = 77.22$. For $c_1 = 20, c_2 = 4.5$, we obtain $\bar{n} = 10.55$ and $v = 47.4$. Thus, in the naive model, goal-directed behavior of individuals improves the system efficiency also in the multiagent case. The reason for this is the underlying assumption that other agents will behave randomly. Using the exact model allows us to drop this assumption.

The exact model. Assume now that agent a in Figure (5.2.a) has the goal to move to square $(2, 2)$, whereas agent b's goal is to move to square $(1, 2)$. We shall analyze how well this type of conflict can be resolved by using the local behavior-based decision algorithms described above. The same question shall be studied by the example shown in Figure (5.2.b) which shows a conflict situation in a shelf corridor that seems (intuitively) harder to resolve.

First, we define the exact model according to Definition 5.2.5. In our example, S is defined by tuples (s_a, s_b). E.g., the initial state is encoded as $s = ((1, 2), (2, 2) \equiv 1222^3$, the goal state is $g = 2212$. Actions in $\mathcal{ACT}$ are pairs (q_a, q_b) denoting simultaneous moves through the grid. Consequently, the transition matrix P defines the probabilities of reaching one (compound) state from another one. The domain theory $\mathcal{D}$ is given by

$$\mathcal{D} = \{ \quad \forall i, j \in \mathcal{ACT} \forall t \in T.loc(i, t) \neq loc(j, t),$$
$$\forall i, j \in \mathcal{ACT} \forall t_i \in T.neighbor(loc(i, t_i), loc(j, t_i))$$
$$\Rightarrow (loc(i, t_{i+1}) \neq loc(j, t_i) \lor loc(j, t_{i+1}) \neq loc(i, t_i) \}$$

where T is a set of time instants, $\mathcal{ACT} = \{1, \ldots, k\}$ a set of agents, $loc :$ $T \times \mathcal{ACT} \mapsto S$, $loc(i, t) = s$ if the location of agent i at time t is s, and $neighbor : S \times S \mapsto [0; 1]$, $neighbor(s_1, s_2) = 1$ iff s_1 and s_2 are neighbors. $\mathcal{D}$ expresses two domain constraints: first, two agents are not allowed to be at the same location at a time; second, agents must not move through each other. Hence, the direct transition from $s = 1222$ to $g = 2212$ is not admissible.

In the following, we compute P using the algorithm described in Definition 5.2.7 based on the individual transition matrices P^a and P^b and domain

---

[3] We shall use this compact notation throughout the example.

theory $\mathcal{D}$. After computing the Cartesian product S' of possible state transitions, the probabilities p'_{ij} for S' are initialized. At this stage, we can restrict our attention to entries (s, s') with $p'(s, s') > 0$. For $s = 1222$, we obtain

$p'(1222, 1112) = \frac{1}{9}$ $p'(1222, 1121) = \frac{1}{36}$ $p'(1222, 1123) = \frac{1}{36}$
$p'(1222, 1312) = \frac{1}{9}$ $p'(1222, 1321) = \frac{1}{36}$ $p'(1222, 1323) = \frac{1}{36}$
$p'(1222, 2212) = \frac{4}{9}$ $p'(1222, 2221) = \frac{1}{9}$ $p'(1222, 2223) = \frac{1}{9}$

Now, S' is divided in admissible and inconsistent states; the only successor of 1222 which is inconsistent with $\mathcal{D}$ is 2212: $IS = \{2212\}$ and $AS = S' - IS$. Next, the compromise states CS are computed from IS. In this example, the only consistent subsequent state when trying to reach 2212 is the initial state 1222, i.e., the local state transitions fail for both a and b. Thus, $\tau(1222, 2212) = \{1222\}$, and $CS = \{1222\}$. $P_{CS}(1222, 1222) = \frac{p'(1222, 2212)}{1} = \frac{4}{9}$. Now, the set of direct successors for 1222 and the corresponding probabilities P are computed for all states. The resulting matrix is of size 36×36. The fundamental matrix N is computed as usual; it can be used to derive the expected number of steps needed to end in an absorbing state given any transient state. Table 5.1 shows the mean number of steps to get from the initial state 1222 to the goal state 2212 for three different initial values for c_1, c_2 in comparison with the optimal solution[4].

		Example (5.2.a)		Example (5.2.b)		Strategy
c_1	c_2	mean # of state trans.	opt.# of state trans.	mean # of state trans.	opt. # of state trans.	
1	1	7.8	3	122.99	8	random
4	1.5	4.86	3	57.4	8	moderately goal-directed
20	4.5	7.91	3	631.48	8	strongly goal-directed

Table 5.1. Mean number of states needed to resolve the conflicts in Figure 3

Surprisingly, the highly goal-directed strategy performs worst for conflict resolution; it is even worse than a random strategy. The moderately goal-directed behavior performs considerably better; it might still be acceptable if we take into account the cost of synchronization in the optimal solution.

Next we analyze the conflict situation shown in Figure (5.2.b). Assuming the goal of agent a is to move from square $(2, 1)$ to $(6, 1)$ and the goal of agent b is to move from $(3, 1)$ to $(1, 1)$, constructing the k-agent Markov chain according to Definition 5.2.7 yields the results shown in Table 5.1. None of the strate-

---

[4] An optimal solution for the example is e.g., $1222 \rightarrow 2221 \rightarrow 2211 \rightarrow 2212$.

gies yields acceptable results[5]. Again, the moderately goal-directed strategy performs best; however, agents the highly goal-directed strategy needs much longer to resolve the conflict than the random strategy. The basic reason for this is that both agents follow a hill-climbing strategy. However, resolving the conflict situation in Figure (5.2.b) requires agent b to make a number of steps in the "wrong" direction, which becomes the more unlikely the more goal-directed b acts.

Discussion. The main result of this analysis is that there are dangerous pitfalls in the local modeling of agents using even very intuitive strategies if these strategies are applied in a multiagent environment. One rash conclusion from this result would be to abolish these strategies and to employ optimal strategies instead, e.g., by coordinating the behavior of the agents globally. However, Latombe [Lat92] showed that global multi-robot path planning is not tractable in general. For this reason, it makes sense to program the behavior-based layer of robots using simple and general algorithms, and then to provide a control architecture that extends these local strategies with the ability to explicitly recognize conflict situations as they occur and to employ other means of conflict resolution in situations where necessary (see Section 4.3).

5.4 Communication vs Local Methods: Empirical Results

In this section, the results of a series of experiments carried out for the loading dock application are reported. The goal of these experiments was to evaluate the behavior of different types of INTERRAP agents and how they depend on different internal and environmental parameters. In particular, it was intended to include those agents with cooperative conflict resolution capabilities in the analysis, and to compare the overall performance of agents that resolve conflict by communication with the performance of other agents who use local methods for conflict resolution.

5.4.1 Agent types

In the following, five exemplary types of INTERRAP agents are defined, that are subject to our empirical analysis.

The random walker (RWK). RWK is an agent that chooses its actions randomly; i.e., it always uses the uniform decision function $\mathcal{F}_u$ (see Definition 4.3.2) for action selection. In the case of RWK, conflict resolution is done implicitly: if the agent selects an alternative that cannot be carried out, execution will fail and the agent will continue selecting alternatives randomly until it has found a solution (if one exists).

---

[5] The state transitions $2131 \rightarrow 3141 \rightarrow 4151 \rightarrow 5152 \rightarrow 6151 \rightarrow 6141 \rightarrow 6131 \rightarrow 6121 \rightarrow 6111$ describe an optimal solution.

Behavior-based agent with built-in conflict resolution (BCR). BCR is a behavior-based agent (i.e., an INTERRAP agent which has neither a local nor a cooperative planning layer. It performs its navigation tasks using a PDF $\mathcal{F}_p$ as described in Definition 4.3.1. Agent BCR is the agent type that has been described by the Markov model in Sections 5.2 and 5.3. In contrast to the random walker, this agent type uses a weighted action selection function that allows it to approach its destination with a high probability.

Behavior-based agent with heuristic conflict resolution (BCH). Similar to BCR, BCH uses decision function $\mathcal{F}_p$ for task planning; however, to resolve blocking conflicts, it uses a reactor PoB as we have described in Section 4.3.4: if possible, it tries to dodge the other agent instead of just moving randomly. Many conflicts can be resolved efficiently by this strategy.

Local planner with heuristic conflict resolution (LCH). LCH uses the hierarchical skeletal planner described in Section 3.5 for local task planning; it employs the same heuristic conflict resolution strategy as BCH.

Local planner with cooperative conflict resolution (LCC). This agent type has the same local planning behavior as LCH; however, for resolving conflicts, it combines conflict resolution via reactor PoBs (for conflicts in hallway and truck regions) with coordination via joint plans (for conflicts in shelf regions).

5.4.2 Description of the experiments

The test series reported in the following contains tests with homogeneous agent societies. We ran experiments with four, eight, and twelve forklift agents. These agents had to carry out randomly generated tasks in a loading dock of size 15 × 20 squares, with six shelves and one truck. The topology of the loading dock (see Figure 4.1) ensures that any square of type *ground* is reachable from any other. The number of tasks were 50 for four agents, 100 for eight agents, and 150 in the twelve-agent case. Each experiment was repeated five times (for twelve agents) and ten times (for eight and four agents), respectively, with the five agent types RWK, BCR, BCH, LCH, and LCC. The focus of the experiment was to evaluate the system behavior with respect to the following questions:

– Is one of the described agent types or conflict resolution strategies dominant for the FORKS application?
– How gracefully degrade the different types and strategies when the number of agents is increased? How robust are they?
– How well do communication-based strategies compared to local ones?

5.4.3 Results

The main results of the experiments are illustrated by the diagrams 5.3.a - 5.3.d.

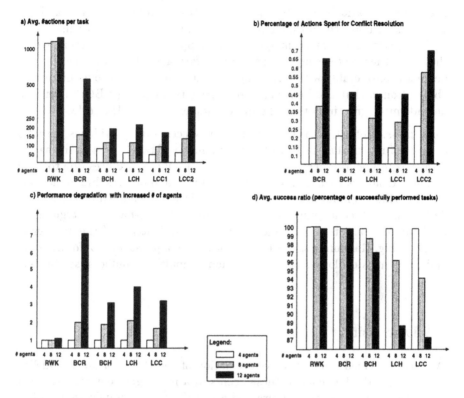

Fig. 5.3. Experimental results for the FORKS application

Absolute performance. Diagram 5.3.a shows the absolute performance for each agent type as the average number of actions needed per task. There are two entries for LCC: LCC1 only accounts for the number of physical actions (moves, turns, gripper actions), whereas LCC2 adds the number of messages sent (one message ≅ one action). RWK performs worst in all experiments. The plan-based types perform better than the behavior-based ones; especially LCC yields the best results in terms of actions; however, the value of explicit coordination depends on the cost of communication.

Conflict Efficiency. Diagram 5.3.b displays the the ratio of actions needed for conflict resolution to the total number of actions. Since RWK does not explicitly recognize conflicts, it is not included in this statistics. The main result to be noted here is that LCC performs well for small agent societies, whereas it actually does not increase conflict resolution efficiency for large agent societies, in comparison with local methods.

Degradation. The factor of performance degradation δ shown in Figure 5.3.c for x agents, $x \in \{4, 8, 12\}$ is computed as $\delta(x) \stackrel{\text{def}}{=} \frac{\#a(x) \cdot \#t(4)}{\#a(4) \cdot \#t(x)} \cdot \frac{1}{\rho}$, where ρ

is the success ratio (see below), $\#a(x)$ denotes the total number of actions, and $\#t(x)$ denotes the total number of tasks in the x-agent experiment.

The performance of agent type RWK happens to be very insensitive to the size of the agent society, whereas the performance of all other agent types degrades considerably with a growing number of agents. A second interesting observation is that the behavior-based agent types (except BCR[6]) tend to degrade more gracefully than the more complex ones (LCH and LCC).

Robustness. Robustness is measured by the success ratio ρ, which is the ratio of successfully finished tasks to the total number of tasks given to the agent. In our experiments, there are three sources of failures. Failures due to local maxima, deadlock situations caused by conflicts, and failures due to multiple conflicts that could not be adequately recognized and handled by the agents. The main result concerning robustness is that behavior-based strategies tend to be more robust than plan-based, cooperative strategies if the number of agents is big. Randomness has been shown to be a powerful tool for avoiding and resolving deadlocks. For a discussion of multiple conflicts, see Section 6.2.

5.5 Discussion

We have provided a quantitative evaluation of the effects of local modeling of agents on the behavior of these agents in a multiagent environment. For the INTERRAP agent architecture and a specific class of applications, a formal model was provided that allows to analyze agents using a specific class of decision algorithms. It both allows us to make quantitative predictions of the performance of different decision algorithms using a strict mathematical framework, and it takes into account two sources of uncertainty: uncertainty in action selection (which is especially important for the evaluation of a multiagent system from the perspective of the external observer) and uncertainty in the execution of actions (which is important from the point of view of the individual agent).

The main question is as to the adequateness of the model: Is it reasonable to assume the Markov property for the decision-making of interacting agents? If asked in this generality, the answer is certainly "no". A major implication of the Markov assumption are that an agent will not change its attitude based on previous experiences, thus its ability to learn is very restricted. However, for the class of algorithms we investigate in this paper, it can be assumed that the agent's decision behavior does not change arbitrarily often over time. Modifications of the behavior caused by learning occur at discrete

---

[6] The poor performance of BCR in the twelve-agent–case is due to a cascade effect resulting from the fact that if there are many other agents around, while trying to resolve a conflict by performing n steps random walk, the agent is very likely to run into a new conflict aso.

points in time. Thus, the theoretical model provided in this work is still useful to explain the behavior of the agent in between changes.

A second point of discussion concerns the ability of the model to deal with more complex forms of agent interaction. Most of the research on agent interaction in DAI [Syc87] [CKLM91] [RZ94] deals with negotiation and requires an agent to maintain a history of the course of the interaction with other agents. Obviously, the Markov property cannot be assumed for these interaction processes, which are located at the local planning and cooperative planning layers of the INTERRAP model. However, we have clearly restricted the scope of the theoretical model to a set of local decision algorithms. Hence, we do not claim to be able to explain cooperative behavior using this model. What we have done to overcome this limitation has been to analyze interaction by communication in the loading dock domain by providing empirical results. A full-fledged theoretical model that allows to analyze and to predict the behavior of INTERRAP agents using more complex types of behavior, such as cooperative planning techniques is beyond the scope of this book (see Section 6.2).

The work reported in this chapter is still at an initial stage in that it only covers local algorithms provided by a particular layer of the INTERRAP architecture, namely the behavior-based one. There are mainly two arguments that justify this research: firstly, our experience with the robotics domain has shown that interactions among robots can be dealt with by local behavior in many cases. Interaction by communication is seldom though vital in some cases. Secondly, as Hanks et al. observed in [HPC93], most other work on evaluating agent architectures has not yet passed the stage of providing empirical results; therefore, we believe that the approach whose initial results we have presented in this chapter is justified.

5.6 Bottom Line

Different conflict handling mechanisms for agents in the loading dock have been evaluated: local, behavior-based mechanisms as well as mechanisms based on communication. It was shown that local mechanisms are useful and often sufficiently effective for the resolution of blocking conflicts in the loading dock. However, coordination by communication and cooperative planning is vital and substantially improves the global performance of the agents in highly constrained situations. Two evaluation techniques have been used: a predictive model based on Markov chains, whose scope is restricted to behavior-based mechanisms based on probabilistic decision functions, and empirical results from a series of experiments with different agent types.

6. Conclusions and Outlook

6.1 Conclusions

The primary claim of this work has been that the extension of a planner–reactor architecture by a cooperation layer is not only possible but also desirable. In Chapter 3, it was shown that such a combination is feasible by defining the layered control architecture INTERRAP. The successful implementation of the FORKS application both as a simulation and as a physical system with real robots, which we described in Chapter 4, evidences the appropriateness of our approach for the design of autonomously interacting agents in real-world domains, and its ability to support the development of robust and flexible multiagent systems.

A secondary hypothesis has been that it makes sense to introduce the notion of procedures at the behavior-based layer of an agent. Procedures are a means for intelligent execution of routine tasks as they incorporate domain knowledge that enables them to deal with a number of disturbing events without causing an explicit replanning process at the local planning layer. This hypothesis has been strongly supported by the fact that we succeeded in building robot agents in the loading dock application with a rather simple planning mechanism (see Section 3.5), yet without shifting too much complexity to the behavior-based layer (see Section 3.4).

The third hypothesis has been that on the one hand, methods on the basis of local behavior-based decision-making play an essential role for interactions among physical systems, but that on the other hand, explicit coordination through communication is of vital importance and can considerably improve the performance of the system in some cases. We have supported this hypothesis in various respects: firstly, the architecture described in this book supports the design of agents and agent interaction both based on local and cooperative methods. Secondly, a simulation testbed has been built that allows to observe different types of interactions among autonomous robots in the loading dock (see Section 4.4). Thirdly, in Section 5.4, we empirically investigated different conflict handling methods, which were either behavior-based or communicative. Fourthly, we presented a probabilistic model allowing us to predict the effects of the modeling of a specific type of behavior-based agents on the over-all performance of the multiagent system (see Chapter 5).

However, this analysis also shows that careful local modeling is crucial for a good performance of the system as a whole, and that even intuitive decisions regarding the design of local agent behavior can have pathological effects if the agent acts in a multiagent environment. Consequently, there is an urgent need for tools for analyzing the interrelationship between the design of agents on the one hand and the design of multiagent systems on the other. This and further open issues shall be discussed in Section 6.2.

In summary, the contribution of this book is a novel control framework for autonomous agents that perform local tasks and that interact with each other in dynamic multiagent domains. This framework supports the reconciliation of reactive and deliberative behavior, and it is geared to the design of agents in multiagent environments, as it provides a variety of interaction mechanisms, from reactive patterns of behavior to negotiation and cooperative planning. The model has been implemented in a rich testbed that we have presented and evaluated by the example of an interacting robots application. We expect that this work should deepen our understanding of a variety of general problems in the design of agent-based systems. It is our hope that the research presented in this book will contribute to the long-term goal of providing robust, flexible, and intelligent control facilities for tomorrow's decentralized, large-scale computer-controlled or computer-assisted systems.

6.2 Future Work

The INTERRAP architecture in its present form has some limitations. In this section, the most important ones are identified, and the possibilities of future research to extend the scope of the architecture are discussed.

6.2.1 Learning and adaptation

The ability of agents to adapt their behavior to changing environmental conditions was mentioned as a necessary prerequisite for intelligence. However, in this work, so far we have neglected the topic of learning. In the following, we shall outline some possibilities of integrating this functionality.

Adaptation of PoBs. The adaptation of the PoBs at the behavior-based layer includes learning new PoBs, modifying existing PoBs, and *forgetting* PoBs that are no longer needed. The generation of new PoBs requires the connection of situations that occur frequently with specific sequences of action suitable to cope with these situations. It is likely for such an extension to start at the local planning layer, as it requires a considerable amount of meta-reasoning.

Another approach to improve the quality of the available PoBs off-line would be to vary the parameters (e.g., individual preconditions) by running a genetic algorithm [Gol89] over the set of PoBs. In a real-robot environment, a large part of the functionality that we modeled at the behavior-based layer might be implemented at the subsymbolic layer, using neural nets that directly link the sensory input of the agent and the control input from higher layers to actoric output. These subsymbolic patterns of behavior could be improved by training the corresponding neural net. See e.g., [Hex96] for an approach to improving the robot's routine behavior.

Finally, forgetting PoBs that are no longer needed requires statistical information about how often a specific PoB has fired. This information has to be weighed against the *importance* of the PoB: the agent should not forget how to deal with a situation that occurs very rarely, but that will have fatal consequences unless handled correctly.

Learning PoBs from plans. PoBs in INTERRAP are conditional plans that are compiled down into procedures which—given an intelligent execution mechanism—make it possible to deal with specific events without explicit replanning. Given a mechanism for planning from first principles that allows us to generate new plans, procedure PoBs could be learned by compiling a set of plans into a procedure. The underlying intuition is to hard-wire routine behavior at a lower layer. Such an extension is supported by the current INTERRAP system as the language used to define the bodies of PoBs (see Section 3.4.2) is almost identical to the plan language used at the local planning layer (see Section 3.5.2).

Learning how to cooperate. Another important source of adaptation relates to the capability of an agent to improve its cooperative behavior. E.g., in an application where there is a market mechanism and negotiation on monetary values, the quality of cooperation can be expressed in terms of the money an agent gains from it. In the loading dock application, we can measure quality of cooperation by how fast conflicts are recognized and resolved. Thus, learning cooperation includes learning new negotiation protocols on the one hand, and improving negotiation strategies, on the other. The former requires the ability to dynamically generate negotiation protocols, the latter requires knowledge about other agents' negotiation strategies; moreover, it requires us to drop the *no history assumption* made throughout this book (see below).

Control aspects. Future work should address the question of *how* to integrate learning into the control architecture. Basically, this can be done either by incorporating the learning capabilities of an agent into a separate control layer located on top of the cooperative planning layer, or by incorporating learning abilities into the individual layers of INTERRAP. The former suggestion implies a global meta layer that monitors and modifies each other control layer. The latter could be generalized to structure each control layer into an object-level and into a meta-level, and to run in the meta-level part what Sloman and Poli have called "meta management processes" in [SP96]. The advantage of such a procedure might be that additional processes and functionalities could be integrated without having to modify the top-level structure of INTERRAP. One possible drawback could be that learning processes that relate to the interplay between the individual layers could be implemented easier from the global perspective of a separate layer.

6.2.2 Communication and negotiation

Extensions of the negotiation framework. The negotiation framework for INTERRAP agents defined in Section 3.6 is based on different assumptions, one of which is the isolated encounter assumption, saying that the negotiation strategy used by an agent to decide which of a set of possible communicative actions it performs next is based on the previous action of the other agent, not on experiences from earlier encounters. Incorporating a history can clearly increase the scope and the effectiveness of agent decisions. However, in doing so, more attention needs to be paid to reciprocity in agent modeling (see e.g., [GD93] [VD96] [Gmy96]).

Currently, another limitation to our work is that while the general negotiation model covers the n–agent case, we have restricted our attention to two-agent encounters as regards the description of interaction in the loading dock application. On the one hand, this has been done to make examples clearer. However, it is a yet unresolved question in how far the concepts described to solve two-agent conflicts by negotiation map to the n–agent case, or: whether a blocking conflict among n agents in the loading dock can be incrementally divided into $n - 1$ independent conflicts between two agents. Whereas this seems possible in some cases, other cases require that agents take their commitments from earlier bilateral negotiations into account when they get involved in new negotiations (see e.g., Figure 4.11 in Section 4.2.4).

Interagent communication language. Currently, the language that is used by agents to communicate with each other is fairly simple (see Section 4.2.2). Establishing a broader communication platform allowing INTERRAP agents to communicate with agents programmed in other agent-based languages is certainly desirable. One way to achieve this goal is to rely on a specific language, and to hope that this language will become a standard to which everybody else adheres. The question of whether e.g., KQML can serve as such a standard being diversely discussed in the scientific community; currently, no other standard is in sight.

A problem of higher-level communication is that there is no semantics that is commonly agreed on for the primitives of the language (speech acts, message types, performatives). Standardizing agent interaction and communication will be infeasible unless there is a standardized definition of the meaning a specific performative has if it is sent to another agent. This is also a prerequisite for an agent to plan communication as a means to achieve its local goals. Thus, the decision of what communication language to choose for INTERRAP agents has to be made in the context of the state of the world-wide effort for standardizing agent communication.

6.2.3 Planning

The focus of this book is on control in a hybrid agent architecture, not on the process of planning itself. Hence future research could be conducted on extending the planning abilities of INTERRAP agents.

Planning from first principles. In Section 3.5, plans were not generated from scratch, but rather selected from a set of pre-stored plans. Whereas the use of predefined plans is legitimate whenever they are available, the integration of a plan generation mechanism from first principles is crucial when dealing with changing environmental conditions; moreover, it facilitates replanning due to its underlying explicit representation of states. There are two bits of recent work aiming at incorporating a planner from first principles into INTERRAP, one of which uses the deductive planner PHI [BBD$^+$92] for the local planning layer of the forklift agents in the loading dock. The second approach is a nonlinear planner that extends the event calculus [KS86] by abduction. The latter approach takes also into account the applicability of the planning mechanism to the generation of multiagent plans.

Local vs cooperative planning. INTERRAP currently restricts the interplay among the local and the cooperative planning layer: the latter is activated by the former only if no plan—or no reasonably good plan—can be found by local planning. On the one hand, this perspective restricts the possibility to look at cooperation as a first-class option to fulfill an agent's local goals. On the other hand, it does not take into account sufficiently that cooperative planning may have to access local planning facilities to evaluate the effects of cooperation and multiagent commitment on the local goals of the agents, and that "doing it oneself" can also be an option in cooperative decision-making. Future work should strive for a more flexible and more expressive interface between both layers by defining additional interaction modes between the two.

Anytime planning. The strict hierarchical control framework of INTERRAP helps avoid several problems of control and coherence; thus, it makes agent programs easier to understand and easier to design. However, it could be of interest to extend our approach to an architecture for *anytime planning*. The basic idea is to have the layers run in parallel and to introduce a global control mechanism which would be a generalization of that presented in this book. Each layer could now generate proposals to deal with specific situations, and the control mechanism could accept the best mechanism. Given a basic solution—which is found quickly by the behavior-based layer—the higher layers could keep on trying to find a better solution while there is still time left. Viewed that way, INTERRAP could be regarded as a discrete anytime architecture producing solutions whose quality monotonously improves over time (see [BD88] [RZ91] [RZ93] [BD94] for related work on anytime algorithms).

However, there are problems that must be solved for such an extension of the control framework: firstly, the required global control mechanism is very complex as the problem of coherence among the decisions of the individual layers will be much harder to solve than it is in INTERRAP. It will probably require the introduction of censor and suppressor mechanisms among the individual control layers, to get rid of which has been one of our research goals. Secondly, the decision of whether there is still time left until some action is required requires an exact model of the requirements of the domain, and is not trivial. Finally, the evaluation of the quality of a solution itself is a hard task in practice, and it requires a considerable research effort.

6.2.4 Evaluation and verification

The design of agents for multiagent environments is a complex task. The human designer of such a system is faced with many design decisions the consequences of which often cannot be clearly seen at the time the agents are designed. Thus, a further line of future research should deal with how the design of agent-based systems can be supported and how existing systems can be analyzed and verified.

Tools. One crucial problem that we discussed in Chapter 5 is the predictability of the global effects of local agent design on the behavior of the multiagent system. The predictive model that we presented to tackle this problem is a beginning rather than a general solution, as its scope is restricted to a specific class of local decision algorithms. More expressive tools are required that allow agent designers to predict the behavior of multiagent systems given descriptions of the individual agents and their interactions. Starting points for such research can be e.g., Burkhard's analytical approach using Petri net models [Bur93], Halpern and Moses' work on distributed knowledge-based system [HM90], and approaches from distributed operating systems [Mat89].

Formalization. A necessary prerequisite for the analysis of agent-based systems are clear formal descriptions of the agents. In Chapter 5, we have given such a formal model for the lowest layer of an INTERRAP agent. An approach to provide a formal basis also for the higher layers of INTERRAP could extend the BDI theory defined in [RG92] by a layered representation of the mental states. The advantage of a formalization of this work is that it may allow us to make further predictions about the behavior of INTERRAP agents and to derive general properties e.g., those referring to the relationship between the informational, motivational, and deliberative state of an agent.

The goal of such an approach should be to obtain general results that show in how far the intuitive design decisions underlying the current form of the architecture (e.g., limiting the structure and the amount of information as well as the reasoning mechanisms at the behavior-based layer to achieve reactivity) have a theoretical justification. However, taking into account the gap that still exists between formal and pragmatic research approaches, the difficulty of such an endeavor must not be underestimated.

6.2.5 Other applications

Applying INTERRAP to further domains is important for measuring its scope and for identifying lacking features. Currently, the agents in the transportation domain (see [FKM94] [FMP95b] [FMP96]), i.e., shipping companies and trucks, are redesigned as INTERRAP-agents. In the transportation domain, the shipping company agents are self-interested utility maximizers, whose goal is to carry out transportation orders at a maximum profit. Within one company, the trucks of that company need to to find good solutions to the problem of scheduling the current set of orders. Thus, the domain provides a rich testbed for the study of game-theoretic concepts, of negotiation protocols for task allocation, and of the relationship between local and cooperative planning. On the other hand, reactivity is required for the truck agents insofar as they have to dynamically adapt their plans to unforeseen environmental conditions such as defects or traffic jams. The question in how far concepts that we have used to model reactive behavior of robots are useful to support this type of reactivity is still unresolved.

INTERRAP shall be applied to several other application domains: one example is the implementation of a MONOPOLY game. The objective of this work is a virtual game environment in which human players interact with INTERRAP agents. Another study investigates a *software agents* domain: a personal assistant for flight management shall be built using the Internet as an operating environment. A third application area envisaged is the problem of the formation of *virtual enterprises* [Mow86]. In this application, the product manager who is responsible for forming the business process describing the virtual enterprise by interaction with the individual participating firms, shall be modeled as an INTERRAP agent.

6.2.6 Knowledge representation

Future work should aim at a further development of the AKB knowledge representation mechanism. In particular, this should include a more expressive query language for the agent knowledge base as well as an extension of AKB by deductive and inferencing capabilities (e.g., for automatically performing knowledge abstraction). The objective of this work is an active knowledge base that runs parallel with the agent control module and that supports the layered model of situation recognition in INTERRAP by providing an active, incremental situation recognizer (see [Mül94] for a preliminary specification of such a mechanism based on a blackboard system). To improve the efficiency of situation recognition, the RETE algorithm shall be used in order to match the activation and monitoring conditions of PoBs against the world model. Finally, in this work, we have neglected belief generation and belief revision. Future research will provide models for them.

A. Proofs of the Theorems

Theorem 3.4.1 (Page 70). *Let t be a point in time, let B be a finite input set of active PoBs; then, the following properties of algorithm MCS hold:*

1. *Algorithm MCS(B) terminates.*
2. *The result B' of algorithm MCS(B) is a maximal compatible set of B, i.e., $mcs(B', B, t)$. As t is fix, we write $mcs(B', B)$.*

Note in the following, that MCS denotes a function call, whereas mcs is a predicate.

Proof: Termination is proven as follows: The sorting procedure terminates as B is finite; termination of the recursive call to MCS is given as in each recursive call in the body of MCS, its the size of its first argument is decreased by one element. Let B_i denote the first argument of the ith recursive call. For $|B_i| = n_i \in \mathbb{N}$, $(\{|B_i| \| i = 1, 2, \ldots\}, <)$ is a Noetherian ordering and termination after $|B|$ steps can be easily proven by induction.

Next, we show that the result of MCS is a maximal compatible subset of B according to Definition 3.4.2. The first condition is satisfied as each $b \in B$ is active. To prove the second part, we introduce two shorthands: firstly, for $k \leq n \in \mathbb{N}$, let B^k denote the set B without its first $k - 1$ elements, i.e., for $B = \{b_1, \ldots, b_n\}$, $B^k \stackrel{\text{def}}{=} \{b_k, \ldots, b_n\}$. Secondly, let $\overline{B^k} \stackrel{\text{def}}{=} B - B^k$.

Before the recursive call of MCS starts, in line 4 of the algorithm, the original set of PoBs B is sorted by descending priority. Now, consider the sequence of recursive calls $MCS(B^1, B'_1)$, $MCS(B^2, B'_2), \ldots, MCS(B^n, B'_n)$, $MCS(B^{n+1}, B'_{n+1})$. In the first part of the theorem, we have shown that for $|B| = n, B = \{b_1, \ldots, b_n\}$, the recursion ends after n steps. Thus, we have $B^1 = B$, $B'_1 = \emptyset$, $B^{n+1} = \emptyset$, and B'_{n+1} is the result set B'.

The main idea of the proof is as follows: for each recursive call $MCS(B^i, B'_i)$, $i \leq n$, we show that the resulting set B'_{i+1} is a maximal compatible subset of $\overline{B^{i+1}}$, i.e., $mcs(B'_{i+1}, \overline{B^{i+1}})$. Thus, especially, for $i = n$, we have $mcs(B'_{n+1}, \overline{B^{n+1}}) \leftrightarrow mcs(B', B - \emptyset) \leftrightarrow mcs(B', B)$, which proves our hypothesis.

The proof is by induction: for $i = 1$, we have $MCS(B^1, B'_1) = MCS(B, \emptyset) = B'_2 = \{b_1\}$. With $\overline{B^2} = \{b_1\}$, we have $mcs(B_2, \overline{B^2})$ which is trivially true.

The induction assumption is that for $1 \leq k \leq n$, $MCS(B^{k-1}, B'_{k-1}) = B'_k$ such that $mcs(B'_k, \overline{B^k})$.

In the induction step, we prove that for $1 \leq k \leq n$, $MCS(B^k, B'_k) = B'_{k+1}$ such that $mcs(B'_{k+1}, \overline{B^{k+1}})$. Consider $MCS(B^k, B'_k)$ with $B^k = \{b_k, \ldots, b_n\}$. Algorithm MCS distinguishes between two cases:

Case 1: compatible(b_k, B'_k). In this case, b_k is incorporated into B'_k: $B'_{k+1} = B'_k \cup \{b_k\}$. We must show that $mcs(B'_k \cup \{b_k\}, \overline{B^{k+1}})$ holds. Because $\overline{B^{k+1}} = \overline{B^k} \cup \{b_k\}$ and due to the induction assumption, this can be reduced to showing that there is no $b' \in \overline{B^k}$ such that $\neg compatible(b_k, b')$ and $b'.priority > b.priority$. Due to the assumption made in the case distinction and due to $B'_k \subseteq \overline{B^k}$, such b' can only be in $\overline{B^k} - B'_k$. However, for $b' \in \overline{B^k} - B'_k$ such that $\neg compatible(b_k, b')$, $b' = b_i, i < k$, and as $i < j \rightarrow b_i.priority \geq b_j.priority$ (because the input list is sorted), therefore $b'.priority \geq b_k.priority$. Therefore, $mcs(B'_k \cup \{b_k\}, \overline{B^{k+1}})$ holds for the first case.

Case 2: $\neg compatible(b_k, B'_k)$. In this case, b_k is not incorporated into B'_k: $B'_{k+1} = B'_k$. Thus, we have $mcs(B'_{k+1}, \overline{B^{k+1}}) \leftrightarrow mcs(B'_k, \overline{B^k} \cup \{b_k\})$. Firstly, due to the induction assumption and $\overline{B^k} \subseteq \overline{B^{k+1}}$, we have $B'_k \subseteq \overline{B^{k+1}}$.

It remains to show that B'_k is a maximal consistent subset of $\overline{B^{k+1}}$, i.e., that for each $b \in B'_k$, there is no $b' \in \overline{B^{k+1}}$ such that $\neg compatible(b, b') \wedge b'.prio > b.prio$.

Obviously, $b_k \in \overline{B^{k+1}} \supseteq B'_k$, and, due to the assumption made in the case distinction, there is $b' \in \overline{B^{k+1}}$ such that $\neg compatible(b_k, b')$. However, as $b' = b_i$ with $i < k$, we have $b.priority \geq b_k.priority$.

Moreover, for $b' \in \overline{B^{k+1}} - \{b_k\} \equiv \overline{B^k}$, and for each $b \in B'_k$, $compatible(b, b')$ holds due to the induction assumption. Therefore, $mcs(B'_{k+1}, \overline{B^{k+1}})$ holds also for the second case. This completes the proof. $\square$

Theorem 3.6.1 (Page 105). *Let N be a finite negotiation set, and let u_i induce a total ordering on N. Then, a negotiation which is carried through using $K, \pi_l, \pi_f, \sigma_1,$ and σ_2 terminates after a finite number of steps.*

Proof: Looking at the finite automaton illustrated in Figure 3.23 representing the negotiation protocol reveals that there is one source of possible non-termination, i.e., a loop of subsequent alternating MODIFY messages. In order to prove termination, we must show that there is no infinite MODIFY loop.

Each time an agent receives a MODIFY message from another agent, it either accepts, in which case the process stops, or it generates a counterproposal and sends a corresponding MODIFY to the other agent. In each MODIFY step, the negotiation set N is replaced by a new negotiation set $N' \subseteq N$. As each agent deletes d upon reception of a MODIFY(d, d') and deletes d' upon sending a message MODIFY(d, d'), it is easy to see that the negotiation sets of both agents are identical after each negotiation step. Due to our assumption, N is finite; thus, N becomes singleton after finitely many steps ($\bar{N} = \{d\}$) if no agreement can be reached before. Assume w.r.o.g. that a_1 is the leader. Then, if $|N| = 2k$, then it is a_2's turn; if $|N| = 2k + 1$, it is a_1's turn. In both cases, the respective agent selects the maximal element from $\bar{N}$ (which is d) and proposes it to the other agent. However, as d is the only remaining element in the negotiation set, it is necessarily the element with the maximal utility; thus, the other agent will agree. Therefore, negotiation terminates. $\square$

Theorem 3.6.2 (Page 114). *A joint plan JP is deadlock-free iff the corresponding graph Gr_{JP} is acyclic.*

Proof: First, assume that JP is deadlock-free. Assume now that the corresponding graph Gr_{JP} contains a cycle $n \rightarrow n_1 \rightarrow n_2 \rightarrow n_k \rightarrow n$. By the construction of Gr_{JP}, for each $n_i \rightarrow n_j$, there exist two plan steps p_i, p_j such that $p_i < p_j$. Thus, there exist plan steps $p, p_1, \ldots, p_k$ with $p < p_1 < \ldots < p_k < p$. As $<$ is transitive, we have $p < p$. This is a contradiction to our assumption that JP is deadlock-free.

The opposite direction of the proof is done analogously: let Gr_{JP} be an acyclic directed graph. Assume now that JP is not deadlock-free, i.e., there are two plan steps p and q such that $p < q$ and $q < p$, i.e., there are plan steps $p \equiv p_1 < \ldots < p_k \equiv q$ and $q \equiv q_1 < \ldots < q_l \equiv p$. By the construction of the corresponding graph, Gr_{JP} contains nodes $n \equiv n_1 \rightarrow \ldots \rightarrow n_k \equiv m$, and $m \equiv m_1 \rightarrow \ldots \rightarrow m_l \equiv n$; thus, there is a cycle $n \overset{*}{\rightarrow} n$ in Gr_{JP}, which is a contradiction to our assumption. $\square$

Theorem 3.6.3 (Page 117). *Let $\mathcal{A} = \{a_1, \ldots, a_n\}$ be a set of agents, Let $\mathcal{JP}$ be the set of joint plans with agents in $\mathcal{A}$. Then, for each $P \in \mathcal{JP}$ and for each a_i, $a_j \in \mathcal{A}$, we have $u_2(a_i, P) = u_2(a_j, P)$.*

Proof: The proof of Theorem 3.6.3 follows directly from $u_2(a, P) = w_2(a, P) - c_2(a, P)$ and from the definition of the global worth and cost functions w_2 and c_2 given in Definitions 3.6.14 and 3.6.11, respectively, which are the same for each agent $a \in \mathcal{A}$. $\square$

Theorem 4.3.1 (Page 160). *Let $\mathcal{F}_u$ be an UDF, let $\mathcal{A} \equiv \{\text{moveto}(d) | d \in \{n, e, s, w\}\}$ be the set of alternatives. Let L be a finite grid of size $n \times m$, let (X_i, Y_i) denote an arbitrary square in L. Then:*

1. *An agent using $\mathcal{F}_u$ as a decision function will reach each square (X, Y) that is reachable from (X_i, Y_i) infinitely often.*
2. *For each $(X, Y) \neq (X_i, Y_i)$ in a nondeterministic scenario, there is no finite upper bound on the maximal number of steps required to reach (X, Y) for the first time.*

Proof: ad 1. The first part of Theorem 4.3.1 follows directly from the random walk theorem stated by Chung [Chu74].
ad 2. Let (X_i, Y_i), $1 \le X_i \le n$, $1 \le Y_i \le m$ be the initial position of the agent. Let $(X, Y) \neq (X_i, Y_i)$, $1 \le X \le n$, $1 \le Y \le m$ be an arbitrary square within grid L. Let $location(s)$ be the access function to the agent's location (X_s, Y_s) in state s.

Assume that there ex. $n \in N$ which is an upper bound of steps required to reach (X, Y) from (X_i, Y_i). This means, for the length $|\alpha|$ of the longest possible sequence of actions $\alpha = (a_1, a_2, \ldots)$, $a_i \in \mathcal{A}$ denoting a sequence of state transitions

$$s_0 \xrightarrow{a_1} s_1 \ldots \xrightarrow{a_n} s_n$$

with $location(s_0) = (X_i, Y_i)$, $location(s_n) = (X, Y)$, and $location(s_i) \neq (X, Y)$ for all $1 \leq i < n$, we have $|\alpha| \leq n$.

Now, we define a sequence β of actions $\{b_1, b_2, \ldots, b_m\}$, $m > n$ and $location(s_0) = (X_i, Y_i)$, $location(s_m) = (X, Y)$, and $location(s_i) \neq (X, Y)$ for all $1 \leq i < m$. We will show that β exists for the set $\mathcal{N} = \{(X_i, Y_i - 1), (X_i, Y_i + 1), (X_i - 1, Y_i), (X_i + 1, Y_i)\}$ of neighbor squares to (X_i, Y_i). This suffices to show that no finite lower bound exists for any other square (X', Y'), since $m' > m$ actions are required to reach (X', Y'). For a direction d, let $\bar{d}$ denote the direction that is opposite to d. We define $\beta = (moveto(d'), moveto(\bar{d}'), \ldots, moveto(d))$, where d' is selected such that the execution of the action $moveto(d')$ does not lead to a state s' with $location(s') = (X, Y)$, where the sequence $moveto(d')$, $moveto(\bar{d}')$ is repeated $\lceil \frac{n}{2} \rceil$ times, and where d denotes the direction corresponding to each $(\hat{X}, \hat{Y}) \in \mathcal{N}$. The assumption that the scenario is nondeterministic guarantees the existence of d'. Obviously, $|\beta| \geq n + 1 > n$.

It remains to show that β is selected with a probability $p(\beta) > 0$. This holds true because $p(\beta) = (\frac{1}{|\mathcal{A}|})^{|\beta|} > 0$. Thus, there is no finite upper bound on the maximal number of steps required to reach (X, Y) for the first time. $\square$

Theorem 5.2.1 (Page 177). *Let M be a finite absorbing Markov chain. Then M^k is a finite absorbing Markov chain.*

Proof: M^k is finite since M is finite and the number of states and the number of actions are identical for M and M^k. To show that M^k is absorbing, we need to show that M^k is of the form

$$\begin{pmatrix} Q' & R' \\ 0' & I' \end{pmatrix}$$

given that M is of the form

$$\begin{pmatrix} Q & R \\ 0 & I \end{pmatrix}$$

By our assumption, M is absorbing. Since M^k differs from M solely by the form of the transition matrix, all we have to show is that the transformation described in Definition 5.2.3 preserves the structure of M. The proof is completed by induction over the rank r of M. For $r = 1$, $M = (1)$ and $M^k = (1)$. M^k is absorbing. Assume now that the assumption is valid for $r \in \mathbb{N}$. Let $rank(M) = r + 1$. Then, M has the form

$$\begin{pmatrix} Q_r & R_r & | & B \\ 0_r & I_r & | & 0 \\ 0\ldots & \ldots 0 & | & 1 \end{pmatrix}$$

M^k has the form

$$\begin{pmatrix} Q'_r & R'_r & | & B' \\ 0'_r & I'_r & | & 0 \\ 0\ldots & \ldots 0 & | & 1 \end{pmatrix}$$

Because of the construction in Definition 5.2.3, we have $\forall k, i, j.P_{ij} = 0 \rightarrow P^k_{ij} = 0$ and $\forall k, i.P_{ii} = 1 \rightarrow P^k_{ii} = 1$. Moreover, by the induction assumption,

$$\begin{pmatrix} Q'_r & R'_r \\ 0'_r & I'_r \end{pmatrix}$$

is an absorbing matrix of rank r. Since there is no restriction on the form of B' in M^k, the transition matrix of M^k can be directly rewritten as

$$\begin{pmatrix} Q_{r+1} & R_{r+1} \\ 0_{r+1} & I_{r+1} \end{pmatrix}$$

Therefore, M^k is absorbing.   □

Theorem 5.2.2 (Page 179). *Let* Ag, M^i, $\mathcal{D}$, *and* $M = (\mathcal{S}, \mathcal{A}, P)$ *be as in Definition 5.2.5 and Definition 5.2.7. Let* M^i *describe a finite absorbing Markov chain for each* $i \in Ag$. *Then* M *describes a finite absorbing Markov chain.*

Proof: To prove that M is an absorbing Markov chain, we need to show that M describes a probability distribution among compound states, and that the transition matrix P can be written in the canonical form

$$\begin{pmatrix} Q & R \\ 0 & I \end{pmatrix}.$$

Let us assume that the number of agents is equal to k and that the local decision matrix of an agent is described by n local states[1]. Then, the number of compound states is equal to n^k. Let in the following, $z \equiv n^k$. To show that P describes a probability distribution, we need to show that for each row i of P (i.e., for each compound state), we have $\sum_{j=1}^{z} p_{ij} = 1$, given the assumption that for each single-agent transition matrix P^l and for each row i in P^l, we have $\sum_{j=1}^{n} P^l_{ij} = 1$.

We shall proceed by proving that this property holds for matrix P' computed in step 2 of Definition 5.2.7, and that the row total is not changed by the operation performed in steps 6 and 7 of the algorithm.

Thus, given k vectors

---

[1] The latter assumption is made as it makes the formulation of the proof more convenient; the extension of the proof for the case that the cardinality of the sets of local states differ is straightforward.

$$\begin{pmatrix} x_{11} & \cdots & x_{1n} \end{pmatrix}$$
$$\cdots$$
$$\begin{pmatrix} x_{k1} & \cdots & x_{kn} \end{pmatrix}$$

denoting the local states of agents a_1 to a_k with

$$\sum_{j=1}^{n} x_{ij} = 1, \qquad (A.1)$$

we have to show that

$$\sum_{\substack{(i_1,\ldots,i_k) \\ \in \{1,\ldots,n\}^k}} x_{1i_1} \cdot x_{2i_2} \cdot \ldots \cdot x_{ki_k} = 1. \qquad (A.2)$$

I.e., the sum of all cross products of the local state probabilities equals one. This is proven by induction over k. We start with $k = 1$. Clearly, we have

$$\sum_{i_1 \in \{1,\ldots,n\}} x_{1i_1} = \sum_{i=1}^{n} x_{1i} = 1$$

which follows from Equation A.1. Now, the induction assumption is stated as follows for $k \in I\!N$:

$$\sum_{\substack{(i_1,\ldots,i_k) \\ \in \{1,\ldots,n\}^k}} x_{1i_1} \cdot x_{2i_2} \cdot \ldots \cdot x_{ki_k} = 1. \qquad (A.3)$$

For the induction step, consider

$$\sum_{\substack{(i_1,\ldots,i_k,i_{k+1}) \\ \in \{1,\ldots,n\}^{k+1}}} x_{1i_1} \cdot x_{2i_2} \cdot \ldots \cdot x_{ki_k} \cdot x_{(k+1)i_{k+1}}. \qquad (A.4)$$

For $1 \le j \le n$, each $x_{(k+1)j}$ is contained in n^k summands of Equation A.4. By factoring out each $x_{(k+1)j}$, we obtain

$$\sum_{\substack{(i_1,\ldots,i_k,i_{k+1}) \\ \in \{1,\ldots,n\}^{k+1}}} x_{1i_1} \cdot x_{2i_2} \cdot \ldots \cdot x_{ki_k} \cdot x_{(k+1)i_{k+1}} \quad =$$

$$x_{(k+1)1} \cdot \sum_{\substack{(i_1,\ldots,i_k) \\ \in \{1,\ldots,n\}^k}} x_{1i_1} \cdot x_{2i_2} \cdot \ldots \cdot x_{ki_k} + \ldots$$

$$+ x_{(k+1)n} \cdot \sum_{\substack{(i_1,\ldots,i_k) \\ \in \{1,\ldots,n\}^k}} x_{1i_1} \cdot x_{2i_2} \cdot \ldots \cdot x_{ki_k} \quad =$$

$$(x_{(k+1)1} + \ldots + x_{(k+1)n}) \cdot \sum_{\substack{(i_1,\ldots,i_k) \\ \in \{1,\ldots,n\}^k}} x_{1i_1} \cdot x_{2i_2} \cdot \ldots \cdot x_{ki_k} \quad =$$

$$\sum_{j=1}^{n} x_{(k+1)j} \cdot \sum_{\substack{(i_1,\ldots,i_k) \\ \in \{1,\ldots,n\}^k}} x_{1i_1} \cdot x_{2i_2} \cdot \ldots \cdot x_{ki_k} \quad \overset{Ind.ass.}{=}$$

$$\sum_{j=1}^{n} x_{(k+1)j} \cdot 1 \quad \overset{Eqn.A.1}{=}$$

$$1 \cdot 1 \quad = 1.$$

This completes the induction proof. Thus, for each row of the matrix P' after the initialization with the cross product of the local probabilities, the row total is equal to one. Hence, step 2 of Definition 5.2.7 results in a probability distribution over the global state transitions.

Next, in step 6, for each state s and each inconsistent state transition $s \rightarrow s'$, the probability $p(s, s')$ is distributed to equal parts over the elements in $\tau(s, s')$; in step 7, it is then set to zero. It is easy to see that each assignment for $p(s, \hat{s}) = p(s, \hat{s}) + \frac{p'(s,s')}{|\tau(s,s')|}$ only changes the row corresponding to s in the transition matrix P. Let the row total of s in P before step 6 be X. Let $|\tau(s, s')| = m$. Then, the new row total X' after steps 6 and 7 is equal to

$$X' = \underbrace{X + m \cdot \frac{p'(s, s')}{m}}_{step6} \underbrace{-p'(s, s')}_{step7} = X.$$

Thus, the reassignment of probability in steps 6–7 does not change the row total as the probability of the inconsistent state transition is assigned to the elements of the row and then set to zero. Thus, matrix P describes a probability distribution.

What remains to show is that the resulting matrix P can be written in the canonical form

$$\begin{pmatrix} Q & R \\ 0 & I \end{pmatrix}$$

of a Markov chain. This can be shown as follows: first, globally absorbing state in M are states $s = (s_1, \ldots, s_k)$ where each s_i, $1 \leq i \leq k$ is a locally absorbing state within M^i. W.r.o.g., let us assume that the single-agent transition matrix has r absorbing and $n - r$ transient states. Thus, M can be sorted such that the up-most and leftmost $n^k - r^k$ rows and columns, respectively denote transient states, whereas the down-most and rightmost r^k rows and columns, respectively, denote absorbing states. Thus, matrix M can be written as

$$\begin{matrix} & \overbrace{\quad n^k-r^k \quad} & \overbrace{\quad r^k \quad} \\ \begin{matrix} n^k - r^k\{ \\ r^k\{ \end{matrix} & \begin{pmatrix} T \rightarrow T & T \rightarrow A \\ A \rightarrow T & A \rightarrow A \end{pmatrix} \end{matrix}$$

There are no transitions from globally absorbing states, neither to any transient state nor to any other globally absorbing state, since by the construction of globally absorbing states from locally absorbing ones, no agent will ever leave a globally absorbing state once it has been entered. Thus, the submatrix $A \rightarrow T$ is a $r^k \times (n^k - n^r)$-dimensional zero-matrix. On the other hand, $A \rightarrow A$ is an identity matrix of rank $r^k \times r^k$. Thus, M can be written in the required canonical form and describes a finite absorbing Markov chain.
□

B. Empirical Results

Table B.1 displays the numerical results of the experiments reported in Section 5.4 with the five agent types in the loading dock. The legend for table B.1 is as follows:

RWK: random walker
BCR behavior-based agent with random conflict resolution strategy
BCH: behavior-based agent with heuristic conflict resolution strategy
LCH: local planner agent with heuristic conflict resolution strategy
LCC: local planner, cooperative conflict resolution strategy
NT: # of tasks
SR: success ratio
NA: # of performed actions
APT: # of actions per task
APC: # of actions per conflict resolution
CRE: % of actions spend for conflict resolution
QDF: quality degradation factor
NM: # of messages sent
MPT: # of messages per task
NMA: Total # of actions + # of messages
MAPT: Actions plus messages per task

APT and MPT have been computed by $\frac{NA}{NO \cdot SR}$ and $\frac{NM}{NO \cdot SR}$, respectively. That means that only successfully finished tasks have been taken into account for computing these values.

The quality degradation factor $QDF(x)$ for an x-agent experiment has been computed by $QDF(x) \stackrel{\text{def}}{=} \frac{NA(x) \cdot NO(4)}{NA(4) \cdot NO(x)} \cdot \frac{100}{SR}$.

Agent Type	RWK			BCR			BCH		
# Agents	4	8	12	4	8	12	4	8	12
NO	50	100	150	50	100	150	50	100	150
SR in %	100	100	100	100	100	100	100	98.6	97.0
NA	53544	107839	166458	3932	15290	84004	3065	11259	27613
QDF	1	1.01	1.04	1	1.94	7.12	1	1.86	3.1
APT	1070.88	1078.39	1109.72	78.63	152.9	560.03	61.3	114.19	189.78
APC	-	-	-	747	5657	54603	644	3941	12702
CRE in %	-	-	-	0.19	0.37	0.65	0.21	0.35	0.46

Agent Type	LCH			LCC		
# Agents	4	8	12	4	8	12
NO	50	100	150	50	100	150
SR in %	100	96.0	88.7	100	94.0	87.3
NA	2697	10611	28948	2541	7528	22032
QDF	1	2.05	4.03	1	1.58	3.31
APT	53.94	110.54	217.57	50.82	80.08	168.25
APC	378	3289	13027	356	2108	9914
CRE in %	0.19	0.31	0.45	0.14	0.28	0.45
NM	-	-	-	356	4704	20052
MPT	-	-	-	7.12	50.04	153.13
NMA	-	-	-	2897	12232	42048
MAPT	-	-	-	57.94	130.13	321.1

Table B.1. Results for homogeneous agent societies of Types RWK, BCR, BCH, LCH, and LCC with 4, 8, and 12 agents

References

[AB70]   J. W. Atkinson and D. A. Birch. *A Dynamic Theory of Action*. John Wiley, New York, 1970.

[AC87]   P. E. Agre and D. Chapman. Pengi: an Implementation of a Theory of Activity. In *Proc. of AAAI-87*, pages 268–272. Morgan Kaufmann, 1987.

[AC90]   P. E. Agre and D. Chapman. What are plans for? In *[Mae90a]*, pages 17–34. 1990.

[Ach05] N. Ach. *Über die Willenstätigkeit und das Denken*. Vandenhoeck und Ruprecht, Göttingen, 1905.

[Ach10] N. Ach. *Über den Willensakt und das Temperament*. Quelle und Meyer, Leipzig, 1910.

[Agh86] Gul A. Agha. *ACTORS: A Model of Concurrent Computation in Distributed Systems*. Series in Artificial Intelligence. The MIT Press, Cambridge, Massachusetts, 1986.

[AHT90] J. F. Allen, J. Hendler, and A. Tate. *Readings in Planning*. Morgan Kaufmann, San Mateo, 1990.

[AIS90] J. A. Ambros-Ingersson and S. Steel. Integrating planning, execution, and monitoring. In *[AHT90]*, pages 735–740. 1990.

[Ark91] R. C. Arkin. Integrating behavioral, perceptual, and world knowledge in reactive navigation. In *[Mae90a]*, pages 105–122. 1991.

[Aus62] J. L. Austin. *How to do Things with Words*. Clarendon, Oxford, UK, 1962.

[Axe84] R. Axelrod. *The Evolution of Cooperation*. Basic Books, 1984.

[BA95]   T. Balch and R. C. Arkin. Motor schema-based formation control for multiagent robot teams. In *Proceedings of the First International Conference on Multiagent Systems*, San Francisco, CA, 1995.

[Bab86] A. Babloyantz. *Molecules, Dynamics and Life. An Introduction to Self-Organization of Matter*. John Wiley and Sons, 1986.

[BBD⁺92] M. Bauer, S. Biundo, D. Dengler, J. Köhler, and G. Paul. Phi - a logic-based tool for intelligent help systems. Technical Report RR-92-52, DFKI, 1992.

[BD88]   M. Boddy and T. Dean. An analysis of time-dependent planning. In *Proceedings of the 7th National Conference on Artificial Intelligence*, pages 49–54, 1988.

[BD94]   M. Boddy and T. L. Dean. Deliberation scheduling for problem solving in time-constrained environments. *Artificial Intelligence*, 67:245–285, 1994.

[BG88]   A. Bond and L. Gasser. *Readings in Distributed Artificial Intelligence*. Morgan Kaufmann, Los Angeles, CA, 1988.

[BHS93]   B. Burmeister, A. Haddadi, and K. Sundermeyer. Generic configurable cooperation protocols for multi-agent systems. In *Pre-Proceedings of MAAMAW93*. University of Neuchâtel, August 1993.

[BIP87]   M. E. Bratman, D. J. Israel, and M. E. Pollack. Toward an architecture for resource-bounded agents. Technical Report CSLI-87-104, Center for the Study of Language and Information, SRI and Stanford University, August 1987.

[BIP88]   M. E. Bratman, D. J. Israel, and M. E. Pollack. Plans and resource-bounded practical reasoning. *Computational Intelligence*, 4(4):349–355, November 1988.

[BKMS96] R. P. Bonasso, D. Kortenkamp, D. P. Miller, and M. Slack. Experiences with an architecture for intelligent, reactive agents. In M. Wooldridge, J. P. Müller, and M. Tambe, editors, *Intelligent Agents — Proceedings of the 1995 Workshop on Agent Theories, Architectures, and Languages (ATAL-95)*, volume 1037 of *Lecture Notes in Artificial Intelligence*, pages 187–202. Springer-Verlag, 1996.

[BM91]   H.-J. Bürckert and H. J. Müller. RATMAN: Rational Agents Testbed for Multi-Agent Networks. In Y. Demazeau and J.-P. Müller, editors, *Decentralized A. I.*, volume 2, pages 217–230. North-Holland, 1991. Also published in the Proceedings of MAAMAW-90.

[Bra87]   M. E. Bratman. *Intentions, Plans, and Practical Reason*. Harvard University Press, 1987.

[Bro86]   Rodney A. Brooks. A robust layered control system for a mobile robot. In *IEEE Journal of Robotics and Automation*, volume RA-2 (1), pages 14–23, April 1986.

[Bro90]   Rodney A. Brooks. A robot that walks: Emergent behaviors from a carefully evolved network. In Patric Henry Winston and Sarah Alexandra Shellard, editors, *Artificial Intelligence at MIT, Expanding Frontiers*, pages 28–39. MIT Press, Cambridge, Massachusets, 1990.

[Bro91]   R. A. Brooks. Intelligence without representation. *Artificial Intelligence*, 47:139–159, 1991.

[BS92]   B. Burmeister and K. Sundermeyer. Cooperative problem-solving guided by intentions and perception. In Y. Demazeau and E. Werner, editors, *Decentralized A. I.*, volume 3. North-Holland, 1992.

[Bur93]   H. D. Burkhard. Liveness and fairness properties in multi-agent systems. In *Proceedings of the 13th Internatinal Joint Conference on Artificial Intelligence (IJCAI-93)*, pages 325–330. Morgan Kaufmann Publishers, Inc., 1993.

[Byl91]   T. Bylander. Complexity results for planning. In *Proceedings of the 12th International Joint Conference on Artificial Intelligence*, pages 274–279, 1991.

[Byl92]   T. Bylander. Complexity results for extended planning. In *Proceedings of the 1st International Conference on Artificial Intelligence Planning Systems (AIPS-92)*, pages 20–27, College Park, MD, USA, June 1992.

[CD96]   B. Chaib-Draa. Interaction between agents in routine, familiar and unfamiliar situations, 1996. to appear in the International Journal of Intelligent and Cooperative Information Systems.

[CDL87]   D. D. Corkill, E. H. Durfee, and V. R. Lesser. Coherent cooperation among communicating problem solvers. In *IEEE Transactions on Computers*, C-36, pages 1275–1291, 1987.

[Chu74]   K. L. Chung. *Elementary Probability Theory with Stochastic Processes*. Springer, New York, 1974.

[CKLM91] S. E. Conry, K. Kuwabara, V. R. Lesser, and R. A. Meyer. Multistage negotiation for distributed constraint satisfaction. *IEEE Transactions on Systems, Man, and Cybernetics (Special Section on DAI)*,
21(6):1462–1477, November/December 1991.

[CL87]   P.R. Cohen and H. Levesque. Persistence, intention and commitment.
In M. Georgeff and A.L. Lansky, editors, *Proceedings of the 1986 Work-
shop on Reasoning About Actions and Plans*, pages 297–340. Morgan
Kaufmann Publishers, 1987.

[CL90]   P.R. Cohen and H.J. Levesque. Intention is choice with commitment.
Artificial Intelligence, 42(3), 1990.

[CP86]   P.R. Cohen and C.P. Perrault. Elements of a plan-based theory of
speech acts. In B.J. Grosz, K. Sparck Jones, and B.L. Webber, editors, *Readings in Natural Language Processing*, pages 423–440. Morgan
Kaufmann Publishers, Inc., Calif., 1986.

[CR79]   E. Chang and R. Roberts. An improved algorithm for decentralized
extrema-finding in circular configurations of processes. *Communiations
of the ACM*, 22:281–283, 1979.

[Dab93]   V. G. Dabija. *Deciding Whether to Plan to React*. PhD thesis, Stanford
University, Department of Computer Science, December 1993.

[Dar59]   C. Darwin. *Origin of Species by Means of Natural Selection*. John
Murray, London, 1859.

[Dar72]   C. Darwin. *The Expression of the Emotion in Man and Animals*. John
Murray, London, 1872.

[DB90]   M. Drummond and J. Bresina. Anytime synthetic projection: Maxi-
mizing the probability of goal satisfaction. In *Proceedings of the Eight
National Conference on Artificial Intelligence (AAAI-90)*, pages 138–
144. AAAI Press / MIT Press, 1990.

[DD92]   A. Drogoul and C. Dubreuil. Eco-problem solving model: Results of the
n-puzzle. In Y. Demazeau and E. Werner, editors, *Decentralized A.I.*,
volume 3, pages 283–295. North-Holland, 1992.

[DDL89]   K. Decker, E. H. Durfee, and V. R. Lesser. Evaluating research in
cooperative distributed problem solving. In L. Gasser and M. N. Huhns,
editors, *Distributed Artificial Intelligence, Volume II*, San Mateo, CA,
1989. Morgan Kaufmann Publishers, Inc.

[Den87]   D. Dennett. *The Intentional Stance*. MIT Press, Cambridge, MA, 1987.

[DKKN93] T. Dean, L. P. Kaelbling, L. P. Kirman, and A. Nicholson. Planning
with deadlines in stochastic domains. In *Proceedings of the Eleventh
National Conference on Artificial Intelligence*, pages 574–579, 1993.

[DL89]   E. H. Durfee and V. R. Lesser. Negotiating task decomposition and
allocation using partial global planning. In *Distributed Artificial Intelligence, Volume II*, pages 229–244, San Mateo, CA, 1989. Morgan
Kaufmann Publishers, Inc.

[DL94]   K. Decker and V. R. Lesser. Designing a family of coordination algorithms. In *Proceedings of the 13th International Workshop on Distributed Artificial Intelligence*, pages 65–84, Lake Quinalt, Washington,
July 1994.

[DLC89]   E.H. Durfee, V.R. Lesser, and D.D. Corkill. Cooperative distributed
problem solving. In Barr et al, editor, *The Handbook of AI*, volume
Vol.4, pages 83–147. Addison Wesley, 1989.

[DR94]   E. H. Durfee and J. Rosenschein. Distributed problem solving and
multiagent systems: Comparisons and examples. In M. Klein, editor,
Proceedings of the 13th International Workshop on DAI, pages 94–104,
Lake Quinalt, WA, 1994.

[DS83]   R. Davis and R. G. Smith. Negotiation as a metaphor for distributed problem solving. *Artificial Intelligence*, 20:63 – 109, 1983.

[DW91]   T. L. Dean and M. P. Wellman. *Planning and Control*. Morgan Kaufmann Publishers, San Mateo CA, 1991.

[EHW+92]   O. Etzioni, S. Hanks, D. Weld, D. Draper, N. Lesh, and M. Williamson. An approach to planning with incomplete information. In *Proceedings of the 3rd International Conference on Principles of Knowledge Representation and Reasoning (KR'92)*. Morgan Kaufmann, 1992.

[ER93a]   E. Ephrati and J. Rosenschein. A framework for the interleaving of execution and planning for dynamic tasks by multiple agents. In K. Ghedira and F. Sprumont, editors, *Preproc. of MAAMAW'93*. Université de Neuchâtel, August 1993.

[ER93b]   E. Ephrati and J. Rosenschein. Multi-agent planning as a dynamic search for social consensus. In *Proc. of IJCAI-93*, pages 423–429. Morgan Kaufmann, San Mateo, CA, August 1993.

[ES89]   E. A. Emerson and J. Srinivasan. Branching time temporal logic. In J. W. de Bakker, W.-P. de Roever, and G. Roezenberg, editors, *Linear Time, Branching Time and Partial Order in Logics and Models for Concurrency*, pages 123–172. Springer-Verlag, Berlin, 1989.

[ES94]   C. Elsaesser and M. G. Slack. Deliberative planning in a robot architecture. In *Proceedings of the AIAA/NASA Conference on Intelligent Robots in Field, Factory, Service, and Space*, 1994.

[Etz93]   O. Etzioni. Intelligence without robots (a reply to brooks). *AI Magazine*, December, 1993.

[EW94]   O. Etzioni and D. Weld. A softbot-based interface to the internet. *Communications of the ACM*, 37(7):72–76, 1994.

[Fer89]   J. Ferber. Eco-problem solving: How to solve a problem by interactions. In *Proceedings of the 9th Workshop on DAI*, pages 113–128, 1989.

[Fer92]   I. A. Ferguson. *TouringMachines: An Architecture for Dynamic, Rational, Mobile Agents*. PhD thesis, Computer Laboratory, University of Cambridge, UK,, 1992.

[Fer95]   I. A. Ferguson. Integrated control and coordinated behaviour. In M. J. Wooldridge and N. R. Jennings, editors, *Intelligent Agents — Theories, Architectures, and Languages*, volume 890 of *Lecture Notes in AI*. Springer, January 1995.

[FF94]   T. Finin and R. Fritzson. KQML — a language and protocol for knowledge and information exchange. In *Proceedings of the 13th Intl. Distributed Artificial Intelligence Workshop*, pages 127–136, Seattle, WA, USA, 1994.

[FHMV95]   R. Fagin, J. Y. Halpern, Y. Moses, and M. Y. Vardi. *Reasoning about knowledge*. MIT Press, 1995.

[FHN71]   R. E. Fikes, P. E. Hart, and N. Nilsson. STRIPS: A New Approach to the Application of Theorem Proving. *Artificial Intelligence*, 2:189–208, 1971.

[Fir89]   R. James Firby. *Adaptive Execution in Dynamic Domains*. PhD thesis, Yale University, Computer Science Department, 1989. Also published as Technical Report YALEU/CSD/RR#672.

[Fir92]   R. James Firby. Building symbolic primitives with continuous control routines. In J. Hendler, editor, *Proceedings of the 1st International Conference on Artificial Intelligence Planning Systems (AIPS-92)*. Morgan Kaufmann Publishers, San Mateo, CA, 1992.

[Fir94]   R. James Firby. Task networks for controlling continuous processes. In *Proceedings of the 2nd International Conference on Artificial Intelligence Planning Systems (AIPS-94)*, pages 49–54, 1994.

[Fis93a]   K. Fischer. The Rule-based Multi-Agent System MAGSY. In *Proceedings of the CKBS'92 Workshop*. Keele University, 1993.

[Fis93b]   K. Fischer. *Verteiltes und kooperatives Planen in einer flexiblen Fertigungsumgebung*. DISKI, Dissertationen zur Künstlichen Intelligenz. infix, 1993.

[FJ91]   J. Ferber and E. Jacopin. The framework of eco-problem solving. In Y. Demazeau and J.-P. Müller, editors, *Decentralized A.I.*, volume 2, pages 181–193. North-Holland, 1991.

[FKM94]   K. Fischer, N. Kuhn, and J. P. Müller. Distributed, knowledge-based, reactive scheduling in the transportation domain. In *Proceedings of the Tenth IEEE Conference on Artificial Intelligence and Applications*, pages 47–53, San Antonio, Texas, March 1994.

[FMP95a]   K. Fischer, J. P. Müller, and M. Pischel. Cooperative transportation scheduling: an application domain for DAI. Research Report RR-95-01, DFKI GmbH, Saarbrücken, 1995.

[FMP95b]   K. Fischer, J. P. Müller, and M. Pischel. A model for cooperative transportation scheduling. In *Proceedings of the 1st International Conference on Multiagent Systems (ICMAS'95)*, pages 109–116, San Francisco, June 1995.

[FMP96]   K. Fischer, J. P. Müller, and M. Pischel. Cooperative transportation scheduling: an application domain for DAI. *Journal of Applied Artificial Intelligence. Special issue on Intelligent Agents*, 10(1), 1996.

[For82]   C.L. Forgy. RETE — a fast algorithm for the many pattern – many object pattern match problem. *Artificial Intelligence*, 19:17–37, 1982.

[Fre52]   S. Freud. *Gesammelte Werke*. Fischer, Frankfurt, 1952.

[Gä88]   P. Gärdenfors. *Knowledge in Flux — Modeling the Dynamics of Epistemic States*. MIT Press, Cambria, Mass., 1988.

[Gas91]   L. Gasser. Social Conceptions of Knowledge and Action: DAI Foundations and Open System Semantics. *Artificial Intelligence*, 47:107–138, 1991.

[Gat91a]   E. Gat. Alfa: a language for programming reactive robotic control systems. In *Proceedings of the IEEE Conference on Robotics and Automation*, 1991.

[Gat91b]   E. Gat. *Reliable Goal-directed Reactive Control for Real-World Autonomous Mobile Robots*. PhD thesis, Virginia Polytechnic and State University, Blacksburg, Virginia, 1991.

[Gat92]   E. Gat. Integrating planning and reacting in a heterogeneous asynchronous architecture for controlling real-world mobile robots. In *Proceedings of AAAI'92*, pages 809–815, 1992.

[GD93]   P. J. Gmytrasiewicz and E. H. Durfee. Reasoning about other agents: Philosophy, theory, and implementation. In *Proceedings of the 12th International Workshop on Distributed Artificial Intelligence*, pages 143–153, Hidden Valley, Pennsylvania, May 1993.

[GeF92]   M. Genesereth and R. Fikes et al. Knowledge interchange format: Version 3.0 reference manual. Technical report, Computer Science Department, Stanford University, 1992.

[Geo83]   M. Georgeff. Communication and interaction in multi-agent plans. In *Proceedings of IJCAI-83*, pages 125–129, Karslruhe, Germany, 1983.

[GH89]   L. Gasser and M.N. Huhns. *Distributed Artificial Intelligence, Volume II*. Research Notes in Artificial Intelligence. Morgan Kaufmann, San Mateo, CA, 1989.

[GI89]   M. P. Georgeff and F. F. Ingrand. Decision-making in embedded reasoning systems. In *Proceedings of the 6th International Joint Conference on Artificial Intelligence*, pages 972–978, 1989.

[GK94]   M. R. Genesereth and S. P. Ketchpel. Software agents. *Communications of the ACM*, 37(7):48–53, 1994.

[GL86]   M. P. Georgeff and A. L. Lansky. Procedural knowledge. In *Proceedings of the IEEE Special Issue on Knowledge Representation*, volume 74, pages 1383–1398, 1986.

[Gmy96]   P. J. Gmytrasiewicz. On reasoning about other agents. In M. Wooldridge, J. P. Müller, and M. Tambe, editors, *Intelligent Agents — Proceedings of the 1995 Workshop on Agent Theories, Architectures, and Languages (ATAL-95)*, volume 1037 of *Lecture Notes in Artificial Intelligence*, pages 143–155. Springer-Verlag, 1996.

[Gol89]   D. E. Goldberg. *Genetic Algorithms in Search, Optimization, and Machine Learning*. Addison-Wesley, 1989.

[GS90]   B. J. Grosz and C. L. Sidner. Plans for discourse. In P.R. Cohen, J.L. Morgan, and M.E. Pollack, editors, *Intentions in Communication*. Bradford Books at MIT Press, 1990.

[Had95]   A. Haddadi. *Reasoning About Cooperation in Agent Systems: A Pragmatic Theory*. PhD thesis, University of Manchester Institute of Science and Technology (UMIST), March 1995.

[Had96]   A. Haddadi. *Communication and Cooperation in Agent Systems: A Pragmatic Theory*, volume 1056 of *Lecture Notes in Artificial Intelligence*. Springer-Verlag, Heidelberg, 1996.

[Hec89]   H. Heckhausen. *Motivation und Handeln*. Springer-Lehrbuch, Berlin, 2nd edition, 1989.

[Hex96]   H. H. Hexmoor. Learning from routines. In M. Wooldridge, J. P. Müller, and M. Tambe, editors, *Intelligent Agents — Proceedings of the 1995 Workshop on Agent Theories, Architectures, and Languages (ATAL-95)*, volume 1037 of *Lecture Notes in Artificial Intelligence*, pages 97–110. Springer-Verlag, 1996.

[HH94]   P. Haddawy and S. Hanks. Utility models for goal-directed decision-theoretic planners, 1994. Submitted to *Artificial Intelligence* journal.

[HM90]   J. Y. Halpern and Y. Moses. Knowledge and common knowledge in a distributed environment. *Journal of the ACM*, 37(3):549–587, 1990.

[HM92]   J. Y. Halpern and Y. Moses. A guide to completeness and complexity for modal logics of knowledge and belief. *Artificial Intelligence*, 54:319–379, 1992.

[HPC93]   S. Hanks, M. E. Pollack, and P. R. Cohen. Benchmarks, test beds, controlled experimentation, and the design of agent architectures. *AI Magazine*, Winter:17–42, 1993.

[HR90]   B. Hayes-Roth. Architectural foundations for real-time performance in intelligent agents. *Real-Time Systems: The International Journal of Time-Critical Computing Systems*, 2:99–125, 1990.

[HR95]   B. Hayes-Roth. An architecture for adaptive intelligent systems. *Artificial Intelligence*, 72:329–365, 1995.

[HRW94]   S. Hanks, S. Russell, and M. P. Wellman, editors. *Working Notes of the AAAI Spring Symposium on Decision-Theoretic Planning*, AAAI Spring Symposium Series, Stanford University, March 1994.

[HSW93]   M. Henz, G. Smolka, and J. Würtz. Oz - a programming language for multi-agent systems. In *Proceedings of the 13th International Joint Conference on Artificial Intelligence*, pages 404–409, Chambery, 1993. Morgan Kaufmann Publishers, Inc.

[Jam90]   W. James. *The Principles of Psychology*. Holt, New York, 1890.

[Jen92a]   N. R. Jennings. *Joint Intentions as a Model of Multi-Agent Cooperation*. PhD thesis, Queen Mary and Westfield College, London, August 1992.

[Jen92b]   N. R. Jennings. Towards a cooperation knowledge level for collaborative problem solving. In *Proceedings of the 10th European Conference on Artificial Intelligence*, pages 224–228, Vienna, 1992.

[Jen94]   N. R. Jennings. *Cooperation in Industrial Multi-agent Systems*, volume 43 of *World Scientific Series in Computer Science*. World Scientific Publishing Co. Inc., 1994. ISBN: 981-02-1652-1.

[K94]   J. Köhler. *Wiederverwendung von Plänen in deduktiven Planungssystemen*, volume 65 of *DISKI - Dissertationen zur KI*. infix, 1994.

[Kae90]   L. P. Kaelbling. An architecture for intelligent reactive systems. In J. Allen, J. Hendler, and A. Tate, editors, *Readings in Planning*, pages 713–728. Morgan Kaufmann, 1990.

[Kan93]   K. Kanazawa, editor. *Working Notes of the Workshop on Dynamically Interacting Robots at IJCAI-93*, Chambery, F, 1993.

[Kau90]   H. Kautz. A circumscriptive theory of plan recognition. In P.R. Cohen, J. Morgan, and M. Pollack, editors, *Intentions in Communication*, pages 105–133. MIT Press, Cambridge, Mass., 1990.

[KG91]   D. Kinny and M. P. Georgeff. Commitment and effectiveness of situated agents. In *Proceedings of the Twelfth Intenrational Joint COnference on Artificial Intelligence (IJCAI-91)*, pages 82–88, Sydney, Australia, 1991.

[Kle91]   M. Klein. Supporting conflict resolution in cooperative design systems. *IEEE Transactions on Systems, Man, and Cybernetics (Special Section on DAI)*, 21(6), November/December 1991.

[KLR+92]   D. Kinny, M. Ljungberg, A. Rao, E. Sonenberg, G. Tidhar, and E. Werner. Planned team activity. In A. Cesta, R. Conte, and M. Miceli, editors, *Pre-Proceedings of MAAMAW'92*, July 1992.

[KNS93]   S. Kraus, M. Nirkhe, and K. P. Sycara. Reaching agreements through argumentation: a logical model (preliminary report). In *Proceedings of the 12th International Workshop on Distributed Artificial Intelligence*, pages 233–247, Hidden Valley, Pennsylvania, May 1993.

[KR89]   M. J. Katz and J. S. Rosenschein. Plans for multiple agents. In L. Gasser and M. N. Huhns, editors, *Distributed Artificial Intelligence*, volume 2. Morgan Kaufmann, 1989.

[KR90]   L. P. Kaelbling and S. J. Rosenschein. Action and planning in embedded agents. In *[Mae90a]*, pages 35–48. 1990.

[KS60]   J. Kemeny and L. Snell. *Finite Markov Chains*. van Nostrand, Princeton, NJ, 1960.

[KS86]   R.A. Kowalski and M.J. Sergot. A logic based calculus of events. *New generation computing*, 4:67–95, 1986.

[KvM91]   T. Kreifelts and F. v. Martial. A negotiation framework for autonomous agents. In Y. Demazeau and J.-P. Müller, editors, *Decentralized A.I.*, volume 2. North-Holland, 1991.

[Lat92]   J. P. Latombe. How to move (physically speaking) in a multi-agent world. In Y. Demazeau and E. Werner, editors, *Decentralized A.I.*, volume 3. North-Holland, 1992.

[Lew35]   K. Lewin. *A dynamic theory of personality: selected papers.* McGraw-Hill, 1935.

[LH92]   D. M. Lyons and A. J. Hendriks. A practical approach to integrating reaction and deliberation. In *Proceedings of the 1st International Conference on AI Planning Systems (AIPS)*, pages 153–162, San Mateo, CA, June 1992. Morgan Kaufmann.

[LS95]   A. Lux and D. D. Steiner. Understanding cooperation: an agent's perspective. In *Proceedings of the First International Conference on Multiagent Systems*, San Francisco, CA, 1995.

[Lux95]   A. Lux. *Kooperative Mensch-Maschine Arbeit – ein Modellierungsansatz und dessen Umsetzung im Rahmen des Systems MEKKA.* PhD thesis, Universität des Saarlandes, Saarbrücken, 1995.

[Mae89]   P. Maes. The dynamics of action selection. In *Proceedings of IJCAI-89*, pages 991–997, Detroit, Michigan, August 1989.

[Mae90a] P. Maes, editor. *Designing Autonomous Agents: Theory and Practice from Biology to Engineering and Back.* MIT/Elsevier, 1990.

[Mae90b] P. Maes. Situated agents can have goals. In *[Mae90a]*, pages 49–70. 1990.

[Mar90]   Frank von Martial. Interactions among autonomous planning agents. In Y. Demazeau and J.-P. Müller, editors, *Decentralized A.I.*, pages 105–119. North-Holland, 1990.

[Mas43]   A. H. Maslow. A theory of human motivation. *Psychological Review*, 50:370–396, 1943.

[Mas54]   A. H. Maslow. *Motivation and Personality.* Harper, New York, 1954.

[Mat89]   F. Mattern. *Verteilte Basisalgorithmen*, volume 228 of *Informatik-Fachbreichte.* Springer-Verlag, 1989.

[Mat93]   M. Mataric. Synthesizing group behaviors. In *Proc. of IJCAI Workshop on Dynamically Interacting Robots*, pages 1–10, Chambery, France, August 1993.

[McC79]   J. McCarthy. Ascribing mental qualities to machines. Technical Report Memo 326, Stanford AI Lab, Stanford, CA, 1979.

[McD08]   W. McDougall. *An Introduction to Social Psychology.* Methuen, London, 1908.

[McD90]   D. McDermott. Planning reactive behaviour: A progress report. In J. Allen, J. Hendler, and A. Tate, editors, *Innovative Approaches to Planning, Scheduling, and Control*, pages 450–458. Morgan Kaufmann, San Mateo, CA, 1990.

[McD91]   D. McDermott. Robot planning. Technical Report 861, Yale University, Department of Computer Science, 1991.

[MFI93]   F. Mondada, E. Franzi, and P. Ienne. Mobile robot miniaturization: A tool for investigation in control algorithms. In *Proc. of the Third Int. Symposium on Experimental Robotics*, Kyoto, Japan, October 1993.

[Min86]   M. Minsky. *The Society of Mind.* Simon and Schuster (Touchstone), 1986.

[MLF96]   J. Mayfield, Y. Labrou, and T. Finin. Evaluating KQML as an agent communication language. In M. Wooldridge, J. P. Müller, and M. Tambe, editors, *Intelligent Agents — Proceedings of the 1995 Workshop on Agent Theories, Architectures, and Languages (ATAL-95)*, volume 1037 of *Lecture Notes in Artificial Intelligence*, pages 347–360. Springer-Verlag, 1996.

[Mow86]   A. Mowshowitz. Social dimensions of office automation. *Advances in Computers*, 25:335–404, 1986.

[MP94]   J. P. Müller and M. Pischel. An architecture for dynamically interacting
            agents. *International Journal of Intelligent and Cooperative Information
            Systems (IJICIS)*, 3(1):25–45, 1994.

[MPT95]   J. P. Müller, M. Pischel, and M. Thiel. Modeling reactive behaviour
            in vertically layered agent architectures. In M. J. Wooldridge and
            N. R. Jennings, editors, *Intelligent Agents — Theories, Architectures,
            and Languages*, volume 890 of *Lecture Notes in AI*. Springer, January
            1995.

[Mül94]   J. P. Müller. A conceptual model of agent interaction. In S. M. Deen,
            editor, *Proceedings of CKBS'94 (Selected Papers)*, pages 213–233. Uni-
            versity of Keele, UK, 1994.

[Mur38]   H. A. Murray. *Explorations in Personality*. Oxford University Press,
            New York, 1938.

[MvdHV91]   J.-J. C. Meyer, W. van der Hoek, and G. A. W. Vreeswijk. Epistemic
            logic for computer science: A tutorial (part one). In *Bulletin of the
            EATCS*, volume 44, pages 242–270. European Association for Theoret-
            ical Computer Science, 1991.

[Neb90]   B. Nebel. *Reasoning and Revision in Hybrid Knowledge Representation
            Systems*. Number 422 in LNAI. Springer, 1990.

[NP77]   G. Nicolis and I. Prigogine. *Self-organization in Non-equilibrium Sys-
            tems*. Wiley Interscience, New York, 1977.

[NS76]   A. Newell and H. A. Simon. Computer science as empirical enquiry:
            Symbols and search. *Communications of the ACM*, 19(3):113–126, 1976.

[Paw27]   I. P. Pawlow. *Conditioned reflexes*. Oxford University Press, London,
            1927.

[Per90]   C.R. Perrault. An application of default logic to speech act theory. In
            P.R. Cohen, J. Morgan, and M. Pollack, editors, *Intentions in Commu-
            nication*, pages 161–185. MIT Press, Cambridge, Mass., 1990.

[RBP91]   J. E. Rumbaugh, M. Blaha, and W. Premerlani. *Object-oriented Mod-
            eling and Design*. Prentice Hall, 1991.

[RG90]   A. S. Rao and M. P. Georgeff. Deliberation and the formation of inten-
            tions. Technical Report 10, Australian AI Institute, Carlton, Australia,
            1990.

[RG91a]   A. S. Rao and M. P. Georgeff. Asymmetry thesis and side-effect prob-
            lems in linear time and branching time intention logics. In *Proc. of the
            12th International Joint Conference on Artificial Intelligence (IJCAI-
            91)*, Sydney, Australia, 1991.

[RG91b]   A. S. Rao and M. P. Georgeff. Modeling Agents Within a BDI-
            Architecture. In R. Fikes and E. Sandewall, editors, *Proc. of the 2rd In-
            ternational Conference on Principles of Knowledge Representation and
            Reasoning (KR'91)*, pages 473–484, Cambridge, Mass., April 1991. Mor-
            gan Kaufmann.

[RG91c]   A. S. Rao and M. P. Georgeff. Modeling rational agents within a BDI-
            architecture. Technical Report 14, Australian AI Institute, Carlton,
            Australia, 1991.

[RG92]   A. S. Rao and M. P. Georgeff. An abstract architecture for rational
            agents. In *Proc. of the 3rd International Conference on Principles
            of Knowledge Representation and Reasoning (KR'92)*, pages 439–449.
            Morgan Kaufmann, October 1992.

[RG95]   A. S. Rao and M. P. Georgeff. BDI-agents: from theory to practice. In
            Proceedings of the First Intl. Conference on Multiagent Systems, San
            Francisco, 1995.

[RMP95] M. Rosinus, J. P. Müller, and M. Pischel. An agent specification language. In *1st International Workshop on Oz Programming*, pages 35–44, Martigny, CH, 1995.

[Ros85] Jeffrey S. Rosenschein. *Rational Interaction: Cooperation among Intelligent Agents*. PhD thesis, Stanford University, 1985.

[Ros96] M. Rosinus. A language for designing INTERRAP agents. Master's thesis, Universität des Saarlandes, Saarbrücken, 1996. Forthcoming.

[RZ91] S. J. Russell and S. Zilberstein. Composing real-time systems. In *Proceedings of IJCAI-91*, pages 212–217. Morgan Kaufmann Publishers, Inc. San Mateo, CA, 1991.

[RZ93] S. J. Russell and S. Zilberstein. Anytime sensing, planning, and action: A practical model for robot control. In *Proceedings of IJCAI'93*, pages 1402–1407, Chambery, F, 1993. Morgan Kaufmann Publishers Inc., San Mateo, CA, USA.

[RZ94] J. S. Rosenschein and G. Zlotkin. *Rules of Encounter: Designing Conventions for Automated Negotiation among Computers*. MIT Press, 1994.

[SA77] R.C. Schank and R.P. Abelson. *Scripts, Plans, Goals, and Understanding*. Hillsdale:Erlbaum, 1977.

[Sac75] Earl D. Sacerdoti. The nonlinear nature of plans. In *IJCAI-75*, pages 206–218, 1975.

[SBKL93] D. D. Steiner, A. Burt, M. Kolb, and Ch. Lerin. The conceptual framework of MAI^{2}L. In *Pre-Proceedings of MAAMAW'93*, Neuchâtel, Switzerland, August 1993.

[Sch89] M. Schoppers. Representation and automatic synthesis of reaction plans. Technical Report UIUCDCS-R-89-1546 (phd-thesis), Dept. of Comuter Science, University of Illinois at Urbana-Champaign, 1989.

[Sea69] J. R. Searle. *Speech Acts*. Cambridge University Press, 1969.

[Sho93] Y. Shoham. Agent-oriented programming. *Artificial Intelligence*, 60:51–92, 1993.

[Sim81] H. A. Simon. *The Sciences of the Artificial*. MIT Press, Cambridge, MA, 2nd edition, 1981.

[Ski35] B. F. Skinner. Two types of a conditional reflex and a pseudotype. *Journal of General Psychology*, 12:66–77, 1935.

[Smi80] R.G. Smith. The contract net protocol: High-level communication and control in a distributed problem solver. In *IEEE Transaction on Computers*, number 12 in C-29, pages 1104–1113, 1980.

[SP96] A. Sloman and R. Poli. SIM_AGENT: A toolkit for exploring agent designs. In M. Wooldridge, J. P. Müller, and M. Tambe, editors, *Intelligent Agents — Proceedings of the 1995 Workshop on Agent Theories, Architectures, and Languages (ATAL-95)*, volume 1037 of *Lecture Notes in Artificial Intelligence*, pages 392–407. Springer-Verlag, 1996.

[ST92] Y. Shoham and M. Tennenholtz. On the synthesis of useful social laws for artificial agent societies. In *Proc. of AAAI-92*, pages 276–281, 1992.

[Ste90] L. Steels. Cooperation between distributed agents through self-organization. In Y. Demazeau and J.-P. Müller, editors, *Decentralized A.I.*, pages 175–196. North-Holland, 1990.

[Ste94] L. Steels. Equilibrium analysis of behaviour systems. In *Proceedings of the 11th European Conference on Artificial Intelligence (ECAI'94)*, pages 714–718, Amsterdam, NL, 1994.

[Syc87] K. P. Sycara. *Resolving Adversarial Conflicts: An approach integrating case-based and analytic methods*. PhD thesis, Georgia Institute of Technology, Atlanta, Georgia, June 1987.

[Tam95]   M. Tambe. Recursive agent and agent-group tracking in a real-time dynamic environment. In *Proceedings of the First International Conference on Multiagent Systems (ICMAS'95*, San Francisco, CA, June 1995.

[Tan88]   A. S. Tanenbaum. *Computer networks*. Prentice Hall Publishers, 2nd edition, 1988.

[Tho98]   E. L. Thorndike. Animal intelligence: an experimental study of associative processes in animals. *Psychological Review Monographs Supplement*, 5:551–553, 1898.

[Tho11]   E. L. Thorndike. *Animal Intelligence*. Macmillan, New York, 1911.

[Tho93]   S. R. Thomas. *PLACA, an Agent Oriented Programming Language*. PhD thesis, Stanford University, 1993. Available as Stanford University Computer Science Department Technical Report STAN-CS-93-1487.

[Tho95]   S. R. Thomas. The PLACA agent programming language. In *[WJ95a]*, pages 355–370. 1995.

[THSS95]   S. Thiébaux, J. Hertzberg, W. Shoaff, and M. Schneider. A stochastic model for actions and plans for anytime planning under uncertainty. *Intl. Journal of Intelligent Systems*, 10(2):155–183, February 1995.

[Tol52]   E. C. Tolman. A cognition motivation model. *Psychological Review*, 59:389–400, 1952.

[TR96]   M. Tambe and P. S. Rosenbloom. Agent tracking in real-time dynamic environments. In M. Wooldridge, J. P. Müller, and M. Tambe, editors, *Intelligent Agents — Proceedings of the 1995 Workshop on Agent Theories, Architectures, and Languages (ATAL-95)*, volume 1037 of *Lecture Notes in Artificial Intelligence*, pages 156–170. Springer-Verlag, 1996.

[Var86]   M. Y. Vardi. On epistemic logic and logical omniscience. In *Proc. of the First Conference on Theoretical Aspects of Reasoning about Knowledge (TARK'86)*, pages 293–306. Morgan Kaufmann Publishers, 1986.

[VD96]   J. M. Vidal and E. H. Durfee. Recursive agent modeling using limited rationality. In M. Wooldridge, J. P. Müller, and M. Tambe, editors, *Intelligent Agents — Proceedings of the 1995 Workshop on Agent Theories, Architectures, and Languages (ATAL-95)*, volume 1037 of *Lecture Notes in Artificial Intelligence*, pages 171–186. Springer-Verlag, 1996.

[vNM44]   J. von Neumann and O. Morgenstern. *Theory of Games and Economic Behavior*. Princeton University Press, Princeton, 1944.

[Wei95]   T. Weiser. Akb: Assertional knowledge base, 1995. Internal Report.

[Wil88]   D. E. Wilkins. *Practical Planning: Extending the Classical AI Planning Paradigm*. Morgan Kaufmann, San Mateo, CA, 1988.

[WJ95a]   M. J. Wooldridge and N. R. Jennings, editors. *Intelligent Agents - Theories, Architectures, and Languages*, volume 890 of *Lecture Notes in Artificial Intelligence*. Springer-Verlag, 1995.

[WJ95b]   M. J. Wooldridge and N. R. Jennings. Intelligent agents: Theory and practice. *Knowledge Engineering Review*, 10(2):115–152, 1995.

[WMT96]   M. J. Wooldridge, J. P. Müller, and M. Tambe, editors. *Intelligent Agents II*, volume 1037 of *Lecture Notes in Artificial Intelligence*. Springer-Verlag, 1996.

[Woo92]   M. J. Wooldridge. *On the Logical Modelling of Computational Multi-Agent Systems*. PhD thesis, UMIST, Department of Computation, Manchester, UK, 1992.

[ZR91]   G. Zlotkin and J. S. Rosenschein. Negotiation and goal relaxation. In Y. Demazeau and J.-P. Müller, editors, *Decentralized A.I.*, volume 2, pages 273–286. North-Holland, 1991.

[ZR93]   G. Zlotkin and J. S. Rosenschein. A domain theory for task-oriented negotiation. In *Proc. of the 13th International Joint Conference on Artificial Intelligence*, volume 1, Chambéry, France, 28.8.–3.9. 1993.

Index

Lecture Notes in Artificial Intelligence
Volumes of Related Interest

Munindar P. Singh
Multiagent Systems: A Theoretical Framework for Intentions, Know-How, and Communications
LNAI 799, 1994. XXIII, 168 pages. ISBN 3-540-58026-3.

Toru Ishida
Parallel, Distributed and Multiagent Production Systems
LNAI 878, 1994. XVII, 166 pages. ISBN 3-540-58698-9.

Michael J. Wooldridge, Nicholas Jennings (Eds.)
Intelligent Agents, ECAI-94 ATAL Proceedings – Second Printing
LNAI 890, 1995. VIII, 407 pages. ISBN 3-540-58855-8.

Michael J. Wooldridge, Jörg P. Müller, Milind Tambe (Eds.)
Intelligent Agents II, IJCAI-95 ATAL Proceedings
LNAI 1037, 1996. XVII, 437 pages. ISBN 3-540-60805-2.

Walter Van de Velde, John W. Perram (Eds.)
Agents Breaking Away, MAAMAW'96 Proceedings
LNAI 1038, 1996. XIV, 232 pages. ISBN 3-540-60852-4.

Gerhard Weiß, Sandip Sen (Eds.)
Adaption and Learning in Multi-Agent Systems,
IJCAI-95 Workshop Proceedings
LNAI 1042, 1996. X, 238 pages. ISBN 3-540-60923-7.

Afsaneh Haddadi
Communication and Cooperation in Agent Systems: A Pragmatic Theory
LNAI 1056, 1996. XIII, 148 pages. ISBN 3-540-61044-8.

John W. Perram, Jean-Pierre Müller (Eds.)
Distributed Software Agents and Applications, MAAMAW'94 Proceedings
LNAI 1069, 1996. VIII, 219 pages. ISBN 3-540-61157-6.

Chengqi Zhang, Dickson Lukose (Eds.)
Distributed Artificial Intelligence: Architecture and Modelling,
1995 Australian DAI Workshop Proceedings
LNAI 1087, 1996. VIII, 232 pages. ISBN 3-540-61314-5.

Lecture Notes in Artificial Intelligence (LNAI)

Lecture Notes in Computer Science